CAVE BEARS AND MODERN HUMAN ORIGINS

The Spatial Taphonomy of Pod Hradem Cave, Czech Republic

Robert H. Gargett

University Press of America, Inc.
Lanham • New York • London

University Press of America,® Inc.
4720 Boston Way
Lanham, Maryland 20706

3 Henrietta Street
London, WC2E 8LU England

Printed in the United States of America
British Cataloging in Publication Information Available

Library of Congress Cataloging-in-Publication Data

Gargett, Robert H.
Cave bears and modern human origins : the spatial taphonomy of Pod Hradem Cave, Czech Republic / Robert H. Gargett.
p. cm.
Includes bibliographical references and index.
1. Pod Hradem Cave (Czech Republic) 2. Paleolithic period--Czech Republic--Brno Region. 3. Mousterian culture--Czech Republic--Brno Region. 4. Neanderthals--Czech Republic--Brno Region. 5. Cave bear--Czech Republic--Brno Region. 6. Animal remains (Archaeology)--Czech Republic--Brno Region. 7. Taphonomy--Czech Republic--Brno Region. 8. Paleontology--Czech Republic--Brno Region--Pleistocene. 9. Brno Region (Czech Republic)--Antiquities.
I. Title.
kGN772.22.C95G37 1996 943.7'02--dc20 95-46518 CIP

ISBN 0-7618-0233-9 (cloth: alk ppr.)

The paper used in this publication meets the minimum requirements of American National Standard for information Sciences—Permanence of Paper for Printed Library Materials, ANSI Z39.48—1984

To my parents

Contents

List of Illustrations

Figure		Page

Figure		Page

Figure		Page

List of Tables

Table		Page

Appendix 2

Table		Page

Acknowledgements

I owe thanks to everyone who encouraged and taught me through the years as a student of Archaeology and Physical Anthropology. I cannot thank them all, but I can justify mentioning the following, beginning with Alan McMillan, Professor of Anthropology at Douglas College in British Columbia, from whom I received my first formal exposure to the discipline of archaeology and early training in excavation.

The Archaeology Department faculty at Simon Fraser University were especially helpful while I was an undergraduate there from 1985-1987. Mark Skinner was always supportive of my formulations of hominid evolution, and provided me with my first paying archaeology job. Brian Hayden hired me as a research assistant, as a field assistant, and as a teaching assistant. Our collaborations were rich and rewarding experiences. He taught me the value of an archaeology that relies on an awareness of the inseparability of theory and empirical observation. He deserves a healthy chunk of my gratitude for his mentorship. Jon Driver's clear presentation style and forthright critical manner will always be a model for me to emulate. And Knut Fladmark's rigorous dedication to the dictum, "natural processes must be ruled out before human effects are inferred," has been perhaps the biggest influence of all.

To my friends during the Simon Fraser years, a special thanks: Glen Guthrie and Chris Knüsel, always ready to debate questions of human origins; Kelly Bush, my spiritual advisor and mentor for much of my time at SFU; Mike Rousseau, the object of my great admiration for his practice of the craft of archaeology.

While I was a doctoral student in UC Berkeley's Department of Anthropology Clark Howell was very supportive of this research, which hopes to contribute to paleoanthropology—the discipline he inspired.

Meg Conkey, my dissertation chair, deserves many thanks. She facilitated this work in numerous ways, and from the beginning expressed her appreciation of its potential contribution and support for its prompt publication. In addition, her dedication to the collegial ideal brought Alison Wylie of the University of Western Ontario's Philosophy Department to Berkeley while I was a student. Wylie's pragmatic approach to science and archaeology was an inspiration. Most importantly, Meg was responsible for my becoming acquainted with Diane Gifford-Gonzalez, of UC Santa Cruz's Board of Anthropology, and through her, better acquainted with animal bones and taphonomy.

Diane has been more than a mentor to me. She suggested that I work on a cave bear assemblage. She also acted as the principal critic of this work, and pointed out to me some of the niceties of reasoning by analogy. Furthermore she suggested the term "spatial taphonomy" in the title. Thanks, Diane.

Lisa Wells, of UC Berkeley's Geography Department, read an earlier draught. I appreciate her input.

I wish to thank the following for their financial support, direct and indirect, of this research: the Regents of the University of California, for making me a Regents Fellow upon entering the graduate program at Berkeley; the Government of Canada, which made me a Doctoral Fellow of the Social Sciences and Humanities Research Council (Award #752-91-00117); the L. S. B. Leakey Fund; the Robert Lowie Fund; the Stahl Endowment.

I also wish to thank the many scholars who responded to my requests for help or information as the project progressed: Emiliano Aguirre, Jesús Altuna, Albert Ammerman, Peter Andrews, Joaquín Arroyo-Cabrales, Anthony D. Barnosky, Gennady Baryshnikov, Hervé Bocherens, C. K. Brain, Peter A. Bull, Jean Clottes, Hillary J. Deacon, F. Delpech, William R. Farrand, Clive Gamble, Russell W. Graham, Donald K. Grayson, John F. Hoffecker, Sebastian Payne, Julian Kerbis-Peterhans, Liora Kolska Horwitz, L. Kordos, Steven L. Kuhn, David J. Meltzer, H. Müller-Beck, Patrick J. Munson, J. S. Oliver, John Parkington, Jean-Philippe Rigaud, Mary C. Stiner, Lawrence G. Straus, Antony Sutcliffe, David Hurst Thomas, Trinidad de Torres

Perez-Hidalgo, Karel Valoch, Paola Villa, Ian Wall, Esmée Webb, William B. White. Esmée Webb suggested that I contact Dr. Valoch in Brno. She is thus in large measure responsible for my having carried out the work presented in this book.

During the early stages of my graduate career Paul Goldberg was helpful in arranging for me to work at Kebara Cave, Israel, with Ofer Bar-Yosef and Bernard Vandermeersch. I am grateful to them for taking such good care of me in the summer of 1989, in spite of our disagreement on the matter of Middle Paleolithic burial. That same summer, with the help of Jean-Philippe Rigaud, I also worked at Roc Allan (Lot-et-Garonne, France) with Alain Turq. It was a most enjoyable three weeks.

Trinidad de Torres Perez-Hidalgo kindly sent me numerous publications on his excellent research on cave bears in the Iberian Peninsula, including his monograph entitled *Osos (Mammalia, Carnivora, Ursidae) del Pleistoceno de la Península Ibérica* (1988). The descriptions and illustrations of the cave bear skeleton were indispensable in the field.

A special thank-you to Keith Kintigh, Andrew Duff, Pete Whitridge, and Colin Grier, for shepherding me through a statistical nightmare.

Appreciation also goes to John Holson of *Pacific Legacy*, in Berkeley, for time on his computer drawing some of the figures in Chapters 3 and 4.

Jerry Rasmussen helped to create some of the computer generated plans in Chapter 7.

Use of a laser printer proved indispensable during the writing and production phase, and would have been a pipe-dream had it not been for the generous bequest of Jeff Lyons, who the world lost to the AIDS epidemic in 1994.

I must thank Gary Haynes for showing me his collection of carnivore modified ungulate bones on visits to Reno, Nevada in 1992 and 1994.

I would especially like to thank Professor Dr. Karel Valoch of the Anthropos Institute, Moravské Museum, Brno, Czech Republic, and Professor Dr. Rudolf Musil of the Masarykovy University, Brno, who made it possible for me to examine the material from Pod hradem Cave excavated by them in the 1950s. Dr. Luděk Seitl, then Director of the Anthropos Institute, deserves thanks for giving me a place to stay while in Budišov. My memories of Zámek Budišov are strong and lasting.

While in Budišov, I found friends whose generosity and good graces made my sojourn there seem at times like a visit to a distant branch of the family. Jiří Zelinka, Jana Dítětová, and her daughter Jana Švaříčková, and Klara Vodová, welcomed me into their homes, gave me presents of baking, invited me for dinners, helped me to find necessaries, kept me company during some dark times, and made sure I was cared for. I will never forget them. I owe them a great deal.

Marshall I. Weisler, my friend and collaborator, is also deserving of mention. His important research into the prehistory of Pacific voyaging and inter-island contact is much emulated by me. And, Marshall commented on the last chapter, for which I thank him.

My family has given me unequivocal support over the years, both moral and pecuniary, and I literally cannot thank them enough. They have never doubted my ability. There could be no more valuable a gift.

This section devoted to thank-yous would not be complete without expressing my love and gratitude to Stephanie Lynn Lambretti. She is last on this list, but her contribution is by no means least. Her unconditional and patient love has buoyed me throughout the often painful process of bringing this project to fulfillment. In addition, her careful and critical reading of this work has improved it. Thanks, Steph.

Finally, I wish it known that none of the people I've thanked are responsible for any errors of fact, or misrepresentations of others' works that may occur in the work before the reader. I take full credit for any *faux pas*, slights, oversights, overstatements, understatements, mistakes, missed chances, blunders, slips, lapses, screw-ups, *errata*, or evidence of textual *naïveté*.

Acknowledgement is due to the following for permission to reproduce illustrations. Figure 1.1 (left): adapted by permission from *American Antiquity*, Volume 52, page 112, Figure 2; © Society for American Archaeology, 1987. Figure 1.1 (right): adapted by permission from *The Pleistocene Old World*, edited by Olga Soffer, page 37; © Plenum Press, 1987. Figure 3.2: adapted by permission from *Die Erforschung der Höhle Pod hradem 1956-1958*; © Moravian Museum, Anthropos Institute, Brno, Czech Republic, 1965. Figure 3.3: adapted by permission from *Die Erforschung der Höhle Pod hradem 1956-1958;* © Moravian Museum, Anthropos Institute, Brno, Czech Republic, 1965. Figure 3.4: adapted by permission from *Die Erforschung der Höhle Pod hradem 1956-1958;* © Moravian Museum, Anthropos Institute, Brno, Czech Republic, 1965. Figure 3.5: adapted by permission from *Die Erforschung der Höhle Pod hradem 1956-1958;* © Moravian Museum, Anthropos Institute, Brno, Czech Republic, 1965. Figure 4.1: adapted by permission from *Journal of Anthropological Archaeology*, Volume 9, page 310, Figure 2; © Academic Press, 1990.

CHAPTER 1

Introduction

This monograph presents a taphonomic and spatial analysis of an Upper Pleistocene cave bear (*Ursus spelaeus*) bone accumulation from Pod hradem Cave, in the Czech Republic (Moravské Museum 1965). The analysis provides evidence pertaining to the question of whether, and to what extent, cave bears and other animals were able to create nonrandom spatial distributions of bone in the cave. The results of this study demonstrate unequivocally that pre-modern hominids and modern humans are not alone in their ability to create patterned horizontal bone distributions in caves. By providing such proof of the ability of non-hominids to create horizontally patterned material traces, I am hoping to foster progress toward a more nuanced understanding of the processes that created spatial patterning in pre-modern hominid sites. This chapter discusses the context within which the research evolved, and my perceptions of the issues and problems surrounding spatial analyses of Middle Paleolithic sites.

Pre-Modern Hominids, Humanness, Culture and Language: Terms, Concepts, and Personal Biases

Before turning to a consideration of the reasons for undertaking this research, some terms will require explanation. "Pre-modern hominid" will refer to archaic *Homo sapiens*, *Homo neanderthalensis*, and skeletally modern fossil forms associated with Middle Paleolithic industries. Overall, theirs is an archaeological signature that does not resemble that of modern humans since the beginning of the Upper Paleolithic (ca. 40 kyr[1] ago). In contrast to the Middle Paleolithic, the Upper Paleolithic is characterized by tool-making techniques and forms that are more readily compared with those known to have been made by modern humans, and which were in use in many places until the modern era.

The dichotomous classification of the Middle and Upper Paleolithic is not universally accepted, nor are its assumptions necessarily self-evident. There is, for example, some question about the meaning of the late Middle Paleolithic or Early Upper Paleolithic Châtelperronian industry in western Europe (see, e.g., Farizy 1990; Mellars 1989) and of the Late Stone Age-like Howieson's Poort assemblages of Middle Stone Age antiquity in Southern Africa (see, e.g., Parkington 1990; Singer and Wymer 1982). Nevertheless, over most of their range, and through considerable time (i.e. about 200 kyr), pre-modern hominids produced a quite uniform archaeological signature that was in many ways unlike that of modern humans. Indeed, this is what all the debate is about, and what leads some paleoanthropologists to question the cognitive abilities of pre-modern hominids in the first place.

"Modern human" will refer to members of the genus *Homo* that are both skeletally and behaviorally modern, associated with Upper Paleolithic and more recent archaeological cultures. I take the term "human" to mean "capable of language and having a culture." I neither assume nor infer on the basis of present evidence that pre-modern hominids were capable of language or culture, as modern humans are. I acknowledge, however, that there is great debate among paleoanthropologists over what language is and how it evolved, and over what culture is, and how it evolved. Indeed, since anthropology was conceived as a discipline, there have been literally hundreds of definitions of culture. Even now, within paleoanthropology, a broad spectrum of viewpoints on the nature of culture is expressed (implicitly or otherwise) among those engaged in research into modern human origins.

Culture

Ideas about culture vary considerably both within Anthropology, and among different national and geopolitical traditions in archaeology and paleoanthropology. Often one's concept of culture is taken for granted, or rarely examined critically. The arguments presented in this chapter, and which underlie the position taken in this book, are based on a view of culture that is not held in common among those working on the question of modern human origins. Furthermore, I would argue that part of the problem of interpretation facing these workers is grounded in differing views of culture that are rarely, if ever, made explicit. This therefore creates ambiguity when those with different construals of culture attempt to debate the significance of archaeological data. For these reasons, it will be useful here to present a brief look at the major views of culture operating in archaeology and paleoanthropology, to ensure that the reader is aware of the particular perspective employed in this work.

On one end of the spectrum, culture is viewed by many (especially American) Paleolithic archaeologists and paleoanthropologists as an adaptive system whereby environmental rigors are overcome through "culture." This is the cultural-materialist framework articulated by Marvin Harris (e.g. 1979). For these scholars, culture and culture change are identified by the technological achievements that "allow" populations to maintain their numbers, ultimately grow, and become more complex (e.g. Clark 1991; Hayden 1993). Social niceties are subsumed under the survival imperative and the technological and economic infrastructure of a society. Culture, then, is seen as the human lineage's "extrasomatic means of adaptation" (White 1959:8, cited in Binford 1965). For Clark, Hayden, and others who share their worldview, there is no question that any stone-tool-making hominids would have had a "culture," because objects outside the biological unit were being fashioned and employed for survival. Language, too, would have evolved in the aid of survival, as a means of maximizing the transmission of information required for efficient exploitation of an environment.

Many European archaeologists cleave to a view of culture that is almost synonymous with ethnicity (see, e.g., Binford and Sabloff 1982; Clark 1991). These scholars conform most closely to the "normative" view of culture that was the object of so much of the New Archaeology's scorn in the 1960s and 70s. They explain interassemblage variability by reference to ideas about how to do things, culturally encoded in the rules of a group. In this view, sets of archaeological traits are used to define cultures, and interassemblage variability is seen as an expression of different "cultures" and their norms. Bordes's contention (e.g. 1961a) that different Middle Paleolithic "cultures" were responsible for Mousterian interassemblage variability is grounded in this view of culture.

My own view of culture is different from that of these and many other paleoanthropologists. I adhere to the idea that culture is a structure of meaning (e.g. Leach 1976), within which technology is only a part, and survival only one of the domains over which culture rules. Language, self-consciousness, symbolic manipulation of material objects, as well as the knowledge required to reproduce a given technology, are all part of culture, all part of being human. This is a view of culture that most closely fits with that espoused by so-called postprocessual archaeologists (e.g. Hodder 1986; Shanks and Tilley 1987a, 1987b, 1989; Wylie 1989b). However, I see this view of culture not so much as a post-anything perspective, but more as a real effort to improve models for understanding past human and hominid behavior (Gargett 1992; Wylie 1989a).

Having this view of culture, I would not call what chimpanzees do "culture," even though they fashion tools from twigs and leaves. The difference between what chimps do and what humans do is the difference between behavior and the thoughts that motivate behavior (an important distinction in humans [Davis 1992]), or the difference between communication and language. The tool, for chimps, is used to get food. The chimp, however, having no structure of meaning like that of humans, has no ability to understand what food means in a cultural sense. Thus, for chimps, there could be no California cuisine, no *Cordon Bleu*. The culturally constructed meaning that Americans attach to food is what enables us to see the difference between hamburgers and *Haute Cuisine*. I do not mean to imply that the Neandertals and their contemporaries were chimpanzees. However, there is presently no consensus as to how nearly human the Neandertals and their contemporaries were. And, with a view of culture such as that just outlined, judging their abilities on the basis of stone tools and bone refuse becomes a more difficult task. For this reason alone it is worthwhile doing research that seeks to facilitate a better understanding of just what cognitive abilities pre-modern hominids possessed.

The arguments for and against the archaeological usefulness of different views of culture are diffuse and legion. This is not the place to review them, nor is it the place to resolve any debate over the nature of culture or of language, or of their evolution. By distinguishing between my view of culture and those of other scholars, and in making my biases explicit in this matter, I hope to have removed what might have been an invisible impediment to the apprehension of my intended meaning. It may be that the adaptationist conception of culture is better suited to the examination of pre-modern hominids than one that includes meaningful cultural constructions, but it may also hamper efforts to "see" any fundamental differences *at the cognitive level*, between pre-modern hominids and modern humans. Clearly, it enables these scholars to see a steady progression of what they see as cultural abilities from the first evidence of stone tool making 2.5 million years ago until the present.

Language

As diverse as concepts of culture are, there are at least as many ideas as to the nature of language, and the animals that belong to the club of the linguistically capable. As it was with concepts of culture, this is not the place to undertake to review the arguments and positions of advocates of this or that idea of language. It is enough to say that the term "language" is used in far too many ways to make it, by itself, a useful criterion for solving questions of modern human origins. With chimpanzees and gorillas capable of making sense in sign language, or with honey bees communicating in their own special dance "language," what separates humans from the rest of the animals becomes a question of the kind of information being communicated. Humans make meaning, and their language communicates that meaning.

The ability to make meaning is not necessary for tool making and survival. Numerous other species get by very well without making meaning, even when they employ objects in their environment to exploit resources to which they would not otherwise have access. Nevertheless, meaning making is an inextricable part of being human and cultural. Most importantly, language is fundamental to being cultural and human. In the final analysis, linguistic ability, meaning, and culture are interdependent (cf. Davidson and Noble 1989). As such, it is difficult to imagine how one of these might have arisen in the absence of the others, or how an archaeological record of a linguistic, meaning making, cultural species would not demonstrate these abilities at some time during its tenure, at some place in its geographic distribution. This is the central lacuna leading researchers like me to question whether or not pre-modern hominids were like modern humans—the archaeology of pre-modern hominids appears virtually static over vast distances and through unimaginable time.

Pre-Modern Hominid Cognitive Abilities

My present position on the question of modern human origins is readily appreciated from reading what I have said thus far, and in other articles and comments on matters of interest to me. A re-examination of Neandertal burial (Gargett 1989a, 1989b), concluded that natural depositional circumstances are as plausible an explanation for the occurrence of the so-called burials as any inferred purposeful behavior on the part of Neandertals. A reading of the archaeological and paleontological evidence convinces me that profound cognitive differences probably existed between pre-modern hominids and early modern humans (Gargett 1990, 1993). These differences seem to be related to the emergence of linguistic, and thus cultural, abilities in the modern humans. Having these views does not stop me, however, from wanting to contribute to a refinement of the archaeologist's ability to explain differences and similarities in the archaeology of Middle and Upper Paleolithic hominids. Nor does it preclude me, ultimately, from changing my mind.

Modern Human Origins: An Interpretive Stalemate?

In undertaking research into the models underpinning spatial analysis of pre-modern hominid archaeological sites, I was motivated by a desire to contribute to a better understanding of modern human origins. As the previous section intimated, there exists deep division among paleoanthropologists over the timing and the character of the emergence of humanness in the hominid lineage. Indeed, the debate has endured since the discovery of the first Neandertal in the 19th century, and has seen the contentious issues vary only slightly in the interim. The last section introduced some of the reasons underlying the debate. Here, the debate will be more fully described.

Biological anthropologists appear to have settled, for the moment, on two mutually exclusive interpretations of the fossil record. One side proposes that skeletally and behaviorally modern humans arose in Africa and spread throughout the world by about 40 to 60 kyr ago, replacing remnant "archaic" populations such as the Neandertals (e.g. Bräuer 1984; Cann et al. 1987; Howells 1976; Stringer and Andrews 1988). The opposing side posits a constant flow of genes among populations of archaic *Homo sapiens*, which resulted in continuous evolution across their vast range, and included clinal variation enabling the perception of morphologically distinct forms in geographically distant populations (e.g. Thorne and Wolpoff 1981; Wolpoff 1980, 1985, 1989; Wolpoff et al. 1984; Wolpoff et al. 1988). Both sides in the debate have historical roots in the frameworks of human evolution posited by Schwalbe, Gorjanovič-Kramberger, Hrdlička, Boule, Howell, and Howells (Smith et al. 1989).

The paleontological picture has recently been complicated by the thermoluminescence dates of 90 to 100 kyr BP (Valladas et al. 1988), and the electron spin resonance date of 115 kyr BP (Schwarcz et al.

1988) obtained for skeletally modern hominid fossils from Qafzeh, in Israel, associated with Middle Paleolithic stone tools. Nearby, at Kebara Cave, a putative Neandertal burial has been thermoluminescence dated at ca. 60 kyr BP (Valladas et al. 1987), and electron spin resonance dated at 60-64 kyr BP (Schwarcz et al. 1989). Thus, skeletally modern hominids and Neandertals were contemporaries for at least 40 kyr, and appear to have behaved identically. Yet, their morphological dissimilarity belies long-separate evolutionary paths. A conundrum, indeed.

Archaeologists are uniquely positioned to infer the behavior and cognitive abilities of pre-modern hominids, because they deal with the products of mental processes, the artifacts. However, archaeologists are no less divided on interpretations of when hominids began to act like modern humans—i.e. when they became capable of language, reflexivity, self-consciousness, symbolic manipulation of material objects, and so on. Some infer that modern cognitive capacities arrived late and relatively suddenly, occurring only in the last 40,000 to 60,000 years, and evidenced, in different places at different times, by the material and other manifestations that attend the so-called Upper Paleolithic—stone tools made in new ways, having forms that continued in use almost to the present; bone and other animal parts worked to fashion subsistence implements as well as items of personal adornment and representational images; colonization of Australia, presupposing modern thought processes necessary to conceive of and effect an ocean crossing of at least 90 km. This group of archaeologists sees language as pivotal to humanness, and inextricably bound up in the production of unequivocal representational images, formalized stone and bone tools, personal ornamentation, and so forth (e.g. Binford 1983, 1985; Chase 1991; Chase and Dibble 1987; Davidson 1991; Davidson and Noble 1989, 1992, 1993; Gargett 1989a, 1990; Klein 1989, 1992; Noble and Davidson 1991, 1993; White 1985). The members of this group would agree with Gargett (1990:244) that "while a transition to modern human behavior did not, after all, coincide with the emergence of modern skeletal morphology, [archaeological evidence suggests that] *it did take place within the modern morphotype*" (italics in original).

Another large group of archaeologists infers that the archaeological record of the Middle Paleolithic is a product of essentially modern humans, or at least humans that had already begun their long journey toward modern humanness hundreds of thousands of years ago—i.e. they were long ago capable of a form of language, manipulation of symbols, of material culture, ritual burial, and so on. This group demonstrates a range of opinion, from those who see no difference between archaic and modern behavior, and thus no difference in cognitive abilities between the two groups, to those who see some difference, but who nevertheless feel that pre-modern hominids were nearly modern humans.

Bar-Yosef et al. (1992:534), for example, in a *Current Anthropology* article on excavations at the important Middle Paleolithic cave site of Kebara, in Israel, see no major differences between the archaeology of the Neandertals and the early modern humans from Europe. Although they explicitly decline to comment on the implications of their findings for the debate on modern human origins, it is clear what their position would be.

In the edited volume *Perspectives on the past*, Clark (1991:108) sums up his feelings about pre-modern hominids by quoting C. Loring Brace's recent epithet, "we have met the neandertals, and they are us!" Given Clark's advocacy of a multiregional origin for modern humans, this makes perfect sense. The Neandertals are, in his view, the last pre-modern hominids before modern humans. In an evolutionary scheme that posits "vectored, clinal change," as well as continuous gene flow across the whole range of pre-modern hominids, Neandertals would need to have been essentially modern to have existed so near in time to modern humans. A more detailed picture of Clark's viewpoint is contained in Lindly and Clark's (1990:239) *Current Anthropology* review of the evidence for symbolic behavior evidenced by morphologically modern hominids during the Middle and Upper Paleolithic. They conclude that there is

> no archaeologically discernible symbolic component [in the pre-modern, skeletally modern hominid archaeological record] ... Does this apparent "fact" render these morphologically modern humans significantly different from or less human than morphological moderns associated with Upper Paleolithic industries in Europe, who admittedly do exhibit symbolic behavior? Are they to be considered outside the evolutionary trajectory of modern humans? Clearly the answer is no.

I would argue that Clark and Lindly's view of culture as an adaptive system allows them to divorce symbolic ability from survival ability, and to see the survival abilities of both pre-modern hominids and modern humans as equivalent, therefore implying cognitive equivalency.

Hayden (1993:140), writing in the *Journal of Human Evolution*, after giving his reading of the archaeological evidence for Neandertal behavior, concludes that

> [in] the European scenario, biological, behavioral and cognitive differences between Middle and Upper Paleolithic populations are slight .... for prehistorians to argue that Neandertals were incapable of religious sentiment, art, burial, foresight, symbolism, sophisticated technology and language is tantamount to invoking ad hominem arguments against many of the best excavators and researchers of this century.

However fervent the conviction, firebrand statements such as this final appeal to authority will not, ultimately, render a false conclusion any less false. If the models used to infer pre-modern hominid cognitive abilities are based on unwarranted premises, the best excavation techniques and analytical skills would be of little use.

Summing up the abilities of the Neandertals in their monograph on the species, Trinkaus and Shipman (1993:418) say that the Neandertals were *nearly* modern. They end by emphasizing that the Neandertals were fully modern in at least three important areas: they buried their dead, they made body ornaments, and they looked after the infirm. Trinkaus and Shipman do not admit that these inferences are supported by equivocal evidence (but see Davidson and Noble 1989; Dettwyler 1991; Gargett 1989a). Believing that these traits characterized the Neandertals, the authors go on to assert that these attributes speak of an awareness of the uniqueness of individuals, and that some may have contributed to the common weal through art, music, or linguistic achievements. Thus, many of the traits associated with the highest ideals in modern humans are attributed to the Neandertals by Trinkaus and Shipman.

As with the scholars from whose work the above examples were drawn, there are a large number of other Paleolithic archaeologists and interested parties who place pre-modern hominids close to or equivalent with modern humans (e.g. Clark and Lindly 1989a, 1989b; Duff et al. 1992; Marshack, e.g. 1989; Schepartz 1993; Stringer and Gamble 1993; Trinkaus and Shipman 1993). Still others are somewhat equivocal on the matter (e.g. Mellars 1989). The arguments and evidence for both sides are voluminous. At issue are interpretations of archaeological evidence, and it is clear that interpretations are in large part colored by whichever view of culture and language is held by the worker. No one group is able to explain, to the satisfaction of all, the differences and similarities between the Middle and Upper Paleolithic, between the behavior of pre-modern hominids and that of modern humans.

I do not believe that another review of the burgeoning archaeological literature dealing with the origins of modern humans is desirable or necessary in this introduction. Numerous major reviews have recently appeared, as well as a number of symposium volumes and semipopular works presenting new evidence and interpretations (e.g. Akazawa et al. 1992; Dibble and Mellars 1992; Dibble and Montet-White 1988; Hayden 1993; Klein 1989, 1992; Lewin 1993; Mellars 1989, 1990; Mellars and Stringer 1989; Nitecki and Nitecki 1994; Rasmussen 1993). The impression one gets from all of these works is that construals of the archaeological record are suspended between two conflicting paradigms—both of which hinge on the inferred or assumed cognitive abilities of pre-modern hominids. It is to this problem of inferring the cognitive abilities of pre-modern hominids that I turn my attention in this work. The remainder of this introduction focuses on some epistemological issues surrounding the history of efforts to infer pre-modern hominid cognitive abilities, especially the role of analogy and actualism (sensu Gifford 1981:367).

Analogical Reasoning and the Problem of Seeing the Present in the Past

There is an inevitable tension between the models archaeologists use to construct knowledge of the past, and the potential for those models to be wrong. This book represents my efforts to enhance the archaeologist's ability to construct strong inferences of past behavior through further specification of the range of possible causal agencies creating spatial patterning.

Models for interpreting the archaeological record are, by necessity, almost always drawn from present experience (see for example Gifford 1981; Gifford-Gonzalez 1989, 1991; Wylie 1982, 1985, 1989a; Young 1989), and depend fundamentally on analogical reasoning (Gifford-Gonzalez 1989, 1993; Wylie 1985). The strength of any model explaining past events relies on the strength of the analogy between phenomena in the present and those of interest in the past. The strongest analogies entail recognition of relevant similarities, which are evidence of similar causal processes operating in the creation of phenomena in the present and those in the past. Wylie (1985) calls these causal processes "determining structures." They are the processual linkages between the past and present phenomena. Construction of strong analogies involves recognizing a modern counterpart (the "source") for the object of interest in the past (the "subject"). Then a search is made in the present for the range of processes that could possibly account for the source phenomenon. Because one has already been satisfied that relevant similarities exist between source and subject, such that the same processes creating the phenomenon in the present could have been responsible for the past phenomenon, the way is clear to infer that the process or processes responsible for the source phenomenon are likely to have been at work creating the subject phenomenon. Thus, by reference to modern examples, i.e. through actualistic research, one infers, by analogy, the causal processes underlying the past phenomenon of interest. This style of reasoning, using relational analogies, is the epistemological underpinning of all research in the middle range, as Binford has characterized it (e.g. Binford 1977, 1981b).

The danger for archaeologists, when employing analogical reasoning, can come in a too-heavy reliance on analogies based on superficial or formal similarities between the modern and the prehistoric context. There is often a temptation to infer too much from formal similarities between what is known in the present and what is observed in the record of the past. One sees formal similarities and from them infers that the same set of functional, mental, social, and other causal factors were involved in the production of the subject. Often these "ampliative" inferences are unexamined for the relations between source and subject that might warrant the use of a particular analogy in the first place.

In spatial analysis, specifically, much is made of the formal similarities between nonrandomly occurring spatial patterns in the past, and the spatial patterns that are deemed to represent cultural behavior in the present. Whether or not there is warrant for the analogy is an empirical question. However, in all investigations of past phenomena the misapplication of a formal analogy can lead to inferences that are simply false.

For example, to answer the question "were pre-modern hominids as intelligent as modern humans?" one might be tempted to compare the size of their brains, because modern humans are smart and have large brains. However, we do not know enough about the determining structures, the causal linkages, between brain size and intelligence, to say if brain size alone is responsible for modern human cognitive abilities. Nor do we know enough about the neurological connection between modern human brains and modern human behavior. Moreover, there is a substantial range of brain sizes in modern humans, with no good evidence to support the claim that there is a one-to-one correspondence between brain size and cognitive ability. Thus, it would be a misuse of a formal analogy to infer, based on brain size alone, that pre-modern hominids and modern humans were equal in their cognitive abilities.

A well-known and uncontroversial archaeological example of the misuse of a formal analogy is that of Dart's (e.g. 1949) hypothesis of an australopithecine "osteodontokeratic technology." During his analysis of Makapansgat, a limestone cave in South Africa, Dart observed that the bone, teeth and horn specimens had been broken in repetitive ways. He concluded that they were tools made by the earliest hominids. He also inferred that fractured baboon skulls were evidence of early hominid hunting. Dart drew these inferences based on what was then current knowledge. As far as was known, the kind of breakage he observed was only attributable to humans. Little was known at the time about cave formation processes, nor was very much known about carnivore habits and their effects on bone. Spurred by doubts about Dart's claims, Brain's (e.g. 1968, 1969, 1974, 1980, 1981) extensive research into the bone collecting and modifying habits of numerous animal species, and into the geomorphology of Makapansgat, have dispelled the myths that grew in the wake of Dart's conclusions.

Brain's work is an excellent example of the use of relational analogy to construct knowledge of past events and behavior. He performed extensive actualistic research into the formation processes of limestone caves, and of the behavior of numerous species of animals with respect to bone. Through his investigations, he arrived at an inferential structure that took account of the causal relationships between the kind of phenomena observed in the archaeological case, and similar phenomena observed in the present. By knowing more about the range of processes that could potentially have produced materials similar to those excavated from the cave, Brain was able to refute Dart's claim that only hominids were capable of producing the bone artifacts from Makapansgat. Indeed, it became apparent through Brain's research that those early hominids had been the hunted, not the hunters.

In summary, actualistic study, the pillar of middle-range research (Binford 1977, 1981b), and one source of strong models for interpreting past events, entails the search in the present for structured, causal or functional relationships, between past and present phenomena. The research in Chapters 2-7 investigates the source side of the inferential equation, hoping to improve the archaeologist's ability to interpret the material traces of pre-modern hominid behavior. In the end, I seek to further specify the range of processes capable of producing spatially patterned material in archaeological sites.

Spatial Analysis and Upper Pleistocene Prehistory: Giving Priority to the Present

It has long been observed that not only the archaeological objects, but also the spatial relations among them, are important in constructing knowledge of past hominid behavior—e.g. vertical relationships have been especially useful in constructing chronologies. Horizontal spatial relationships in Paleolithic sites, however, have not always been monitored with the same degree of precision as they are in most archaeological excavations today. The horizontal dimension has of course been considered important (implicitly or otherwise) in analyses of sites with permanent architecture, and which therefore have a prominent horizontal structure. However, distributions of archaeological remains range from the highly organized and unmistakable structures of ancient cities such as Pompeii to diffuse and unstructured scatters of the debris of stone tool manufacture on the Great Plains. What concepts and assumptions underpin the spatial analysis of these less obviously structured artifact distributions?

The idea that patterning is a necessary outcome of human activity seems to play a large part in the framework of interpreting spatial patterning. Humans such as ourselves, it is thought, behave in patterned ways. Any patterned material traces of human behaviors, then, must themselves reflect the behavioral patterns that led to their existence (see e.g. Hill 1966:10; Schiffer 1983). Although some modification by natural physical and biological processes is expectable, the hope is that one should be able to diagnose behavior from artifact spatial patterns. With few exceptions (e.g. Binford 1983; Kroll and Isaac 1984) most of the literature on spatial analysis has been formulated explicitly or otherwise in relation to modern humans. Problems arise when such models are applied to the behavior of pre-modern hominids without first examining whether or not the perceived similarities between modern human and pre-modern hominid spatial patterning are the result of similar underlying processes.

As evidence of the almost universal expectation that patterned (modern) *human* behavior results in a spatially patterned archaeological record, I offer the following sample of rare explicit statements of the assumptions underlying spatial analysis and the determinants of spatial patterning. Although these authors had modern humans in mind when they made these statements, depending on one's assumptions about the "cultural" abilities of pre-modern hominids, such statements can and do inform the analysis of pre-modern hominid spatial patterning to a greater or lesser extent. As a final note regarding the following statements, given what has already been said in this introduction about the variety of conceptions of culture, the following must be viewed more as examples of a set of assumptions about the expectation of patterning, than as examples of particular ways of interpreting that patterning in a cultural framework.

Clarke (1977:10) defines spatial structures as "...the nonrandom output of *human choice processes* which allocate structural forms, activities and artefacts to relative loci within sites ..." (italics mine). For Simek and Larick (1983:165) the analysis of spatial distributions involves an "ambition to relate form [i.e. pattern] to process [i.e. *cultural behavior* resulting in the pattern]" (italics mine). In a discussion of the variability in interhousehold spacing among foragers, "site structure" for Gould and Yellen (1987:77) is a "...[reflection of] the existence of *culture-specific* norms..." (italics mine). Little (1985:34) treats all pattern in an archaeological context as the result of the "behavior of the people who consciously and unconsciously deposit their *material culture*" (italics mine). Earle (1985:77) is explicit, although tentative, in his introduction to the special issue of *American Archeology* on pattern recognition: "Cultural behavior is patterned, and so too must be the artifactual detritus resulting from that behavior; such is our faith." Furthermore, archaeologists must "...use that patterning to reconstruct past *cultural* behavior" (italics mine).

Leroi-Gourhan (1950), a pioneer of wide-area excavations, believed that there is a cultural message written into the archaeological record. He warned that archaeologists must be careful to keep track of the spatial relationships of everything so as not to overlook any potentially important messages. Leroi-Gourhan held that unconscious deep structures in the human mind, akin to the structures, or grammar, of a language, manifested themselves in behavior that left material traces—and might therefore be observed in the spatial patterning of archaeological sites. Leroi-Gourhan (1950:6) devised an elegant analogy to illustrate his point. Imagine a layered cake in which the words "Happy Birthday" are inscribed, not on the top, as is customary, but between the layers. If the cake is cut into small pieces (i.e. if the archaeological site is excavated in horizontally limited vertical excavation units), the message could easily go unrecognized. Thus he viewed it as important to employ wide-area excavations, and three-dimensional point provenience of all artifacts.

From my reading of the literature, it appears that spatial analyses of sites without permanent architecture are driven by the expectation that activities—e.g. butchery, hide preparation, cooking—tend to occur in confined spaces (i.e. "activity areas"). In addition, tools associated with those activities (i.e. "tool kits") will be discarded in the activity area (for a thorough discussion of the rationale for spatial analysis, see Carr [1984]; Binford et al. [1970], Whallon [1973a, 1973b, 1974]). Refuse disposal is also believed to follow rules, whether defined by the particular culture or by the physical parameters of the refuse and the living space (e.g. Binford 1978b; Hayden and Cannon 1983; O'Connell 1987).

There has been much discussion lately about whether or not expectations of spatial integrity at Paleolithic sites are defensible (e.g. O'Connell 1987). It seems that the spatial segregation of activities in modern hunter-gatherer camps is often not as clear as archaeologists would have wanted, nor do such activities always map onto identical activity areas in successive occupations of the same site. These are significant questions, given the tendency for most Paleolithic sites to consist of a palimpsest of occupations. However, there still seems to be reason to hope that, at least in some circumstances, such as at Pincevent, France (Leroi-Gourhan and Brézillon 1966, 1972), spatial patterning can be used to infer complex human behavior (see, e.g., Enloe 1991, 1992; Enloe and David 1989; Gargett and Hayden 1991).

In summary, intrasite spatial analyses of modern human archaeological sites are freighted with assumptions about the structured relationship between patterned modern human behavior (including culture) and spatial patterning. Regardless of the definition of culture adhered to by any given researcher, such models can and do find their way into spatial analyses of pre-modern hominid sites. Despite the historic reliance on knowledge of modern human behavior for interpreting spatial patterning created by hominids whose behaviors may not have been motivated by modern human cognitive abilities, to date there has been no explicit attention paid to the epistemological issues raised by this practice. It is an empirical question whether or not models of modern human use of space can be applied to that of pre-modern hominids.

Some Spatial Analyses of Middle Paleolithic Sites

The last section gave several examples of the theoretical connection between expectations of modern human behavior and spatial patterning. Although none of the authors cited was concerned explicitly with problems pertaining to pre-modern hominids, similar theoretical formulations may be seen in the work of a number of researchers whose subject is the behavior of pre-modern hominids. To illustrate the apparent ease with which models of modern human behavior are employed to describe and explain prehistoric spatial patterning, the following examples are offered. Two of the most widely publicized "spatial analyses" from the Paleolithic date from an earlier period than is the focus of this study. However, the received view of the "habitations" at Lazaret Cave (de Lumley 1969a), and at the site of Terra Amata (de Lumley 1969b), both in France, has led to the perception that complex, culturally based spatial patterns were in evidence long before the period of interest in the present work. De Lumley inferred, from the presence of what appear to be nonrandom clusters of artifacts and some putative post moulds, that a commodious beach hut, composed of bent saplings, enclosed a social unit. In addition, a series of what were thought of as discrete depositional units contained what appeared to be roughly similar arrangements. A thorough re-examination of both sites and the behavioral information they contain is not within the purview of this research. However, there is reason to be circumspect about de Lumley's inferences in both cases.

Villa (1982), after conjoining a sample of the lithics from Terra Amata, concluded that the horizontal patterning observed by de Lumley may not be, in fact, the aggregate of contemporary objects. The spatial patterns observed by de Lumley were based on what he called living floors, inferred from their apparent stratigraphic integrity. However, Villa found that many conjoinable pieces came from different stratigraphic positions, separated by as much as 20 to 30 cm. Villa (1982:285) thus cautions that "reconstructions [of the lifeways of the Terra Amata hunters] should be considered as largely speculative because they are based on ambiguous or inadequate data."

Lazaret Cave has been the subject of a monograph (de Lumley (1969a) and a film (de Lumley 1971). A generation of archaeologists has grown up with the image of a wolf skull left behind by the hominid occupants, symbolically barring the door to the "hut" inferred to have been constructed in the cave. In addition to the symbolic content implied by the wolf skull, de Lumley has made much of the co-occurring spatial patterns that are interpreted as spatial divisions within the hut. Unfortunately the evidence for the hut itself is scanty, unless one accepts that the spatial patterning "proves" the existence of a hut inside the cave.

In both of the above cases no warrant is ever offered as to why the spatial patterns are inferred to mean what de Lumley says they mean. Indeed, this is the essential shortcoming of all of the examples given here, and of the method of spatial analysis, in general, as it has been applied to prehistoric human or hominid contexts. Perhaps this book will shed some light on ways of viewing the patterns discerned at Lazaret and elsewhere.

Turning now to the Middle Paleolithic, at L'Hortus Cave, in the south of France, De Lumley (1972) painstakingly piece-plotted every object of interest in this fissure fill. In the monograph, every level is presented in plan, with every artifact and bone indicated, and there are numerous cross sections. The authors find few potentially meaningful concentrations of bone, and make much of "empty" spaces, which they tentatively attribute to the presence of skins placed on the floor, which would have collected debris and, once removed, left a void. This is a glaring example of the misplaced use of a formal analogy. Nothing warrants the inference, from a space free of artifacts, of skins placed on the floor. That is simply one, very speculative, possibility.

Villa (1982:285) also mentions l'Hortus in her discussion of vertical displacement and spatial integrity. She notes that the

> remains of a single animal (a wolf) have been found scattered through eight separate layers with a vertical spread of about 1 m. In four instances teeth and jaw fragments belonging to the same Neanderthal individuals (Hortus 2, 3, 5, 9) were scattered through vertical distances of about 50 cm, across four or five layers.

Thus, even if de Lumley had discovered interpretable patterning, it would be difficult to uncritically accept the inferences drawn from them.

Payne (1983:161), in his study of the fauna from Douara Cave, in Syria, uses visual inspection of the bone distributions to distinguish those introduced by owls and bats from those introduced by hominids. He finds the bones introduced by Middle Paleolithic hominids to be "scattered," as opposed to those introduced by the animals, which he finds to be "clustered," due to their habit of roosting and creating spatially constrained accumulations of their own bones and those of prey species. Payne offers no explanation for the "scattered" nature of the distributions left by hominids.

Hietala and Stevens (1977:539) investigate spatial patterning at Rosh Ein Mor, an open-air site with Mousterian stone tools in the Negev Desert, in Israel. They demonstrate that the distribution and association of several categories of stone tools is nonrandom. They interpret this patterning as potential evidence of sedentism, presumably because they infer that the assemblage accumulated over a period of time, and that any perceptible spatial patterning must therefore be evidence of persistent occupation.

The excavators of the Mousterian deposits in Kebara Cave (Bar-Yosef et al. 1992) delineate several concentrations of bone. There are areas of accumulation near the north wall, time-transgressive circular aggregates away from the walls, and wide areas where no bone appears. The heavy concentrations against the north wall are considered to be dumping zones; the circular areas, intentional arrangements; large areas with few bone specimens are considered to have been cleared of bones.

Rigaud and Geneste (1988) interpret three overlapping concentrations of lithic, faunal and other artifacts as evidence of something they call "social cohesion" among the Mousterian groups occupying couche VIII at Grotte Vaufrey, in France. In a subsequent study, Rigaud and Simek (1991), employing a *K*-means spatial analysis on the same materials at Grotte Vaufrey, are able to quantitatively identify several co-occurring concentrations of artifactual material. These authors are circumspect, and do not ascribe any more significance to these clusters than to say that they may represent preferred areas of the cave. Rigaud and Simek also explicitly acknowledge the possibility that other animals may have contributed to the patterning, mostly through disturbance of patterning that existed before their arrival.

Otte et al. (1988), in their investigation of the Middle Paleolithic site of La Grotte de Sclayn, in Belgium, were unable to identify any meaningful patterning in the stone and bone. Although they publish density maps, which display concentrations of artifacts, the authors explicitly decline to say if these were telling them anything about the behavior of the hominids.

Bonifay (1962, 1964) observed a number of anomalous stone arrangements in La Grotte de Regourdou, near Montignac, in France. The superficial similarity to "walls," "coffers," and "cairns" led him to claim that these had been purposely constructed by Neandertals. In my re-examination (Gargett 1989a) of the stone "structures," it became apparent that well-understood physical processes, such as cave breakdown, had been ignored in favor of elaborate inferences of Neandertal ritual and burial.

All of these authors, it seems, are seeking or seeing patterns they can relate to present experience. However, present experience is not a good teacher if there is insufficient warrant to claim that similar processes led to the creation of the phenomenon of interest. As long as researchers resort to only one framework, i.e. the ethnographic present, for interpretation of spatial patterning associated with pre-modern hominids, those interpretations will be susceptible to considerable error and thus, to radical revision.

Actualistic Research and Spatial Analysis: Prospects for Inferring Pre-Modern Hominid Cognition

The preceding sections of this introductory chapter have laid out the epistemological basis for interpretations of past hominid behavior, and suggested some shortcomings of the present approach to spatial analysis. Specifically, not enough attention has been paid to the potential for non-culturally mediated animal behavior to pattern the material in a prehistoric space. The solution suggested for improving the interpretive strength of such analyses involved the search for processes in the present that would be capable of producing the past phenomenon of interest. Although such actualistic research nearly always involves applying knowledge of present processes, gained in the present, carrying out such studies in fact avoids the trap of simply imposing present categories on past phenomena. It does this by recognizing relevant similarities between past and present phenomena, such that causal linkages can be securely inferred, and which strengthen the analogy between past subject and present source. This section discusses research aimed at understanding the processes responsible for patterning observed in the material traces of past behavior.

In what has until now been a largely implicit inferential framework, the meaning of spatial patterns in pre-modern hominid sites seems to be infused with notions about what creates spatial patterning in modern human archaeological sites. These may be notions about cultural rules having to do with where activities should take place (even the idea that there are distinct things called "activities" that the actors understand as such). They may also be notions about where refuse should go (even the notion that there are categories such as "refuse" in the heads of the hominids).

In order to advance the debate about the emergence of the pre-modern hominid capacity for culture I take the advice of Binford (1977, 1981b), Gifford-Gonzalez (1989, 1991), and Schiffer (1972, 1983, 1987): archaeologists need to continually acquire a broader base from which to understand the processes (including mental) structuring all archaeological sites (i.e. the material traces, such as artifacts, ecofacts, and spatial patterns). This will involve more, not less, research aimed at warranting the linkages, assumed or inferred, between processes observable in the present and the products of similar processes observed in the archaeological domain. It will also require warranting inferences about the cognitive abilities of the hominids themselves. In other words, it will involve middle-range research, both of the physical traces, and of the mind (for a discussion of the prospects for a middle-range theory of the mind, see Cowgill 1993). The research presented here may help in achieving that end, through investigating the efficacy of spatial analysis in understanding pre-modern hominid behavior.

There have been numerous efforts of late to "get at" the behavior of archaic hominids through new lines of evidence, as well as through new ways of looking at old evidence. There have been many actualistic and comparative studies of human and animal modification of bone aimed at understanding the processes that formed the archaeological record of both behaviorally modern and archaic hominids. These studies comprise the wider context for my research into the abilities of non-hominids to produce spatially patterned material in their cave occupations.

Ethnoarchaeological studies, such as that of the Nunamiut by Binford (1978a), and actualistic studies of bone-modifying animals, such as wolves, bears, elephants (Haynes 1980, 1982, 1983a, 1983b, 1985, 1988a, 1988b), hyenas (Bunn 1983; Blumenschine 1986, 1987, 1988, 1989, 1991; Hill 1976, 1977, 1979, 1980, 1983), and modern humans (Binford 1978a, 1978b, 1981b), more recent studies of the treatment of killed and scavenged carcasses by the Hadza (O'Connell et al. 1988; Bunn et al. 1988), as well as theoretical examinations of the assumptions governing such studies (e.g. Gifford-Gonzalez 1993; Kent 1993), have attempted to sort out the question of the kind of decision making manifested in bone assemblages. Most recently, a kind of paleo-actualism has been employed, similar to that used in this research. Stiner's (1990a, 1990b, 1991a, 1991b, 1992, 1993, 1994) careful comparative study involved use of "controls" drawn from fossil bone accumulations known to have been produced by Pleistocene carnivores, to provide the inferential basis for predator/prey relations and the pre-modern hominid niche in the Upper Pleistocene of western central Italy.

Animals and Space: A Gap in Spatial Analysis Research

Potentially damaging to the current methodological foundation for investigating spatial patterning in Middle Paleolithic archaeological assemblages is the possibility that other animals, not possessing culture or acting in ways that we would call human, might also be capable of creating patterns in the refuse comprising their own "archaeological" sites. Not much attention has been paid to any spatial patterning that might result. Although Koby (1941, 1953), Kurtén

(1976), and Schmid (1976) have all commented on the presence of apparent spatial patterning in fossil bone accumulations, with the exception of a small-scale plan in Schmid's article, these reports were anecdotal. More recently, Kerbis-Peterhans (1990) and Lam (1992) observed spatial patterning in and around hyena dens. However, theirs were not systematic studies aimed at understanding the causes and consequences of spatial patterning, *per se*. I submit that the rarity of such investigations results from a widespread assumption that only humans are capable of creating behaviorally meaningful nonrandom distributions of bones and stones.

The central problem in interpreting hominid spatial patterning without acknowledging the potential for other animals to produce patterning is an epistemological one, involving the use of a misplaced formal analogy. If it can be shown that non-hominids, as well as pre-modern hominids and modern humans, can produce nonrandom spatial patterns, there would be no necessity to invoke modern human cognitive abilities or culturally mediated behavior of any kind to explain the presence of spatial patterning in pre-modern hominid archaeological sites (assuming, of course, that this is always a warranted inference where modern human spatial patterning is concerned). Such a finding would encourage a more nuanced interpretation of pre-modern hominid spatial patterning, by removing the one-to-one relationship that now exists between the expectation that spatial patterning reflects modern human behavior, and the observation of patterning in the residues of pre-modern hominids.

Reconsidering Pre-Modern Hominid Cognition: Some Encouraging Results from Lithic Studies

Finally, by way of demonstrating the value of investigating the warrant for analogies used in the interpretation of pre-modern hominid cognitive abilities, this section contains illustrations of how recent examinations of Middle Paleolithic technology have revealed shaky reasoning in the study of pre-modern hominid stone tool making. In so doing these studies have called into question long-held beliefs about the cognitive abilities of pre-modern hominids.

Both examples given below fundamentally question the inference that certain lithic "types" are desired end-products, requiring long sequences of flake removal and core preparation, and thus representing a high degree of cognitive sophistication akin to that of modern humans. One example, having to do with Mousterian stone tool variability, has particular importance in the debate over when modern humans emerged, because one particularly persuasive way of viewing the artifacts seems to have structured all interpretations of Upper Pleistocene pre-modern hominid cognitive abilities for fifty years or more. In the received view, Middle Paleolithic stone tool diversity and interassemblage variability have come to be seen as an indication of the cultural sophistication of the Neandertals and other pre-modern hominids. It now appears, however, as if the diversity of Mousterian stone tool types has been exaggerated. Moreover, the source of interassemblage variability may have at least as much to do with the limiting factors of raw material availability and climate, as with any inferred cultural response. It might therefore be argued that Mousterian hominid cognitive abilities have been concomitantly overestimated.

François Bordes (1961b), in defining and codifying "types" in Mousterian chipped stone tools from France, imposed present categories on past behavior, and in so doing, unwittingly diverted scholarly attention from the real nature of Mousterian interassemblage variability. Bordes (1961b) defined 63 stone tool types, 21 types of handaxes, and later (1980) added 9 unique Levallois flakes (and preparations) to round out the Middle Paleolithic repertoire. As a result, generations of scholars have been acculturated to the notion that Mousterian diversity was equal in many ways to the diversity observed in unequivocally modern human, i.e. Upper Paleolithic, assemblages (see, e.g., Simek and Price 1990).

Mousterian stone workers are believed to have made dozens of different tool types, from "side scrapers" to "burins" to "handaxes." According to the dominant paradigm, each was made to a design—a mental template—in the mind of the pre-modern hominid. This belief has had the greatest impact on Middle Paleolithic studies. It is a belief born of a dominant twentieth century conceptual framework.

Bordes began, as many would, by recognizing modalities in the chipped stone artifacts of the Mousterian—they could be sorted into classes of objects that were manufactured in a similar fashion and had ostensibly similar shapes. In the mind of a modern stone-knapper, and in the minds of archaeologists, such modalities can and do represent emic "types," possessing a particular function and, perhaps, some arbitrarily imposed style—e.g. "end scraper," "projectile point," "knife," "Desert side-notch," "Alberta point," "Clovis Point," and so on. Bordes, like most archaeologists of his day or any other, quite naturally assumed that the modalities he observed—i.e. the Mousterian types he defined—were objects having similar functions, the shapes of which were decided upon in advance by their makers.

Bordes (1953) also observed that the Mousterian "types" occurred in varying frequencies from site to site, and from level to level within sites. These assemblages, too, demonstrated seemingly exclusive modalities, and were in turn accorded the status of types or "Mousterian facies"—i.e. the Mousterian of Acheulean Tradition, the Denticulate Mousterian, the Typical Mousterian, the Quina Mousterian. Bordes's Mousterian facies definitions, and the epistemological framework underpinning them, have dominated thinking about the Middle Paleolithic for decades. Even the Binfords, who disagreed with Bordes on the interpretation of Mousterian interassemblage variability (Binford, S. R. 1971; L. R. Binford and S. R. Binford 1966; S. R. Binford and L. R. Binford 1969), originally accepted the types as real end products of desired shape and composition, and thus, that they represented tools having different functions.

In a series of articles Dibble (1984, 1985a, 1985b, 1987a, 1987b) argues persuasively that many of Bordes's types, rather than being desired end products, are in fact arbitrary stopping points along a continuum of reduction from flake blank through discarded nub. Bordes's types are thus like snapshots of a time-transgressive phenomenon involving successive episodes of retouch. In illustrating his thesis, Dibble (1987a:112, figure 2) demonstrates how a transverse scraper results from continued reduction of a flake that began as a single, lateral-edge scraper (see Figure 1.1). In another, Dibble (1987b:37, figure 1) depicts the sequence that leads to definition of at least three Bordesian types: 1) single scraper, 2) bi-convex double scraper, 3) convex convergent scraper (see Figure 1.1). These illustrations demonstrate the quality of arbitrariness in Bordes's type definitions. At points along the continuum, loss or discard resulted in the tool's description as a different one of Bordes's types. Dibble's conclusions are robust. Thus, a number of Bordesian scraper types will likely come to be viewed not as the result of premeditation, but as the product of repeated retouch for the purpose of maintaining a sharp-edged flake. This weakens Bordes's contention a) that pre-modern hominids made stone tools according to predetermined ideas about how a stone tool should turn out, and b) that Middle Paleolithic hominids made a great variety of stone tools. Although the conclusion to be reached regarding pre-modern hominid cognitive ability is by no means straightforward, Dibble's work will no doubt continue to have an increasing impact on the debate about the modernness of pre-modern hominids.

In addition to contributing to the deconstruction of Bordes's Mousterian scraper typology, Dibble and Rolland (1992), and Rolland

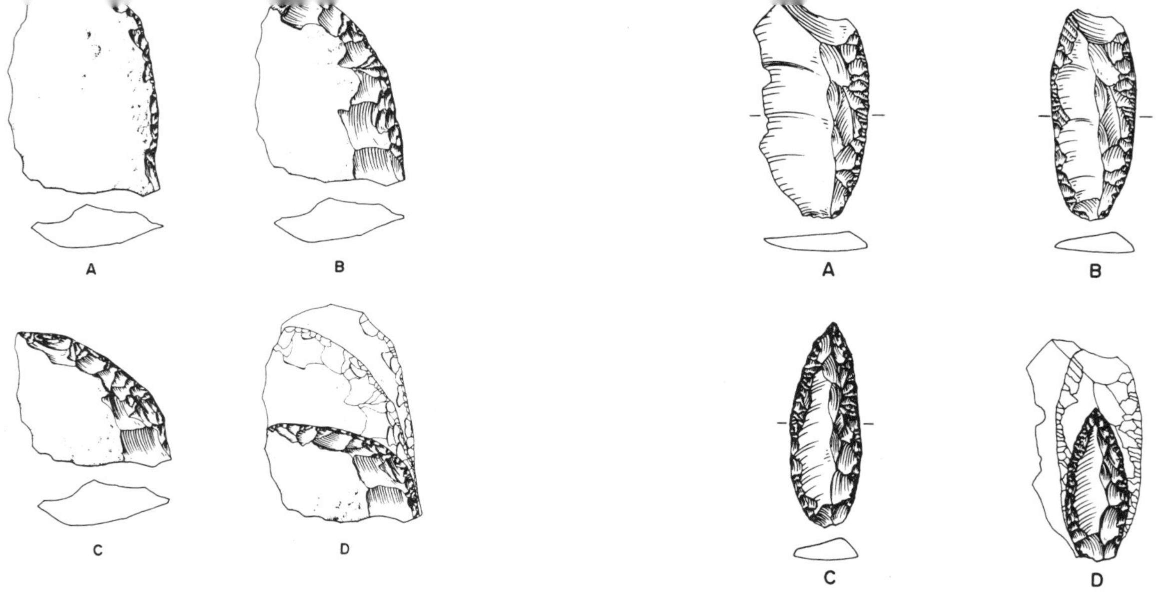

Figure 1.1. Sequences of Mousterian flake tool reduction. Left, how continued reduction of a flake that began as a single, lateral-edge scraper, A, results in a transverse scraper, C (after Dibble 1987a:112, figure 2). Right, the sequence that leads to definition of at least three Bordesian types: A, single scraper; B, bi-convex double scraper; C, convex convergent scraper (after Dibble 1987b:37, figure 1).

and Dibble (1990) have argued that much of the variability in Mousterian *assemblages* is the result, not of the ideas or cultural norms in the minds of pre-modern hominids, but of the number of times a given flake was retouched, which in turn seems to have been a function of proximity to, and availability of, raw material. Thus Mousterian assemblage variability need not be viewed, as Bordes (1961a) contended, as the result of cultural traditions persisting through time, and varying according to the geographic distribution of, for example, the Mousterian of Acheulean Tradition. Nor is variability necessarily to be seen as the result of different site functions, as the Binfords originally posited (Binford, S. R. 1971; Binford, S. R. and Binford, L. R. 1969).

As Dibble (1987a) explains, both the tool "types" and the Mousterian "facies" are still useful categories for monitoring the stone-tool-making behavior of pre-modern hominids, and as a common language for the discussion of what variability exists across the vast range of those same hominids. What all this means for interpretation of pre-modern hominid cognitive abilities is not at all clear. However, it was this kind of rethinking that emboldened me to question assumptions about spatial analysis, and to undertake the research presented in this book.

Indeed, it would seem as if the time is ripe for re-examinations of prevailing notions about pre-modern hominid cognition, although as yet those who openly question whether pre-modern hominids had language or culture are few (e.g. Binford 1984, 1985; Davidson 1991; Davidson and Noble 1989, 1992, 1993; Gargett 1989a, 1989b, 1990, 1991, 1993; Noble and Davidson 1991, 1993). In another construal of the lithic evidence for pre-modern hominid cognitive abilities, Davidson and Noble (1993) argue that to say mental templates were responsible for the shapes of Middle Paleolithic stone tools is exemplary of a wider underlying phenomenon in paleoanthropology, what they call the "finished artifact fallacy." Davidson and Noble go a step further than Dibble, in suggesting that both the handaxe and the Levallois technique for making flakes of predetermined shape—both considered hallmarks of the presumed sophistication and premeditative ability of hominids in the Middle Paleolithic—are not so. Rather, they too are the products of unwarranted presumptions, such as Bordes's, regarding Mousterian types (cf. Barton 1990).

Davidson and Noble (1993) argue that the handaxe, for example, has long been viewed as evidence of a "mental template" in the minds of the hominids that made them, because it has been perceived as having been made to a standardized shape. They argue that it is simply an elongated bifacial core, the shape of which is determined more by the shape of the flake on which it is produced, than by premeditation on the part of its maker. In a similar vein, they see Levallois flakes as the result of a time-transgressive reduction phenomenon, not as preconceived end products, which they have historically been taken to be. The elaborate technique of platform preparation observed on proximal Levallois flakes, Davidson and Noble see as a necessary precursor to removal of a large flake from the convex surface of a core, once the edge angles on the core were such that it could no longer be used to produce large, useful flakes. If the flake removal was profound enough, the core became useful again, and was further reduced. This would explain why the Levallois core, with its characteristic Levallois flake scar, is a common feature of Middle Paleolithic assemblages: if the final flake removal is not profound, the core cannot be used further and is discarded; if it is, the core can be further reduced and a Levallois core cannot thus be discovered and typed by archaeologists.

I do not mean to malign Bordes, or anyone else, for the Middle Paleolithic interpretive framework developed over the past century, and which the work of Dibble and Rolland has called into question. I simply wished to illustrate that misplaced formal analogies to modern human behavior can confuse interpretations of the past. It might be prudent, after all, to start from the premise that pre-modern hominids were different from modern hominids, notwithstanding the abundant (and by no means widely accepted) arguments in favor of pre-modern hominid sophistication based on brain endocasts (e.g. Falk 1983; Holloway, e.g. 1974, 1985), or on equivocal "art" (Marshack, e.g. 1989), or on presumed Piagetian stages evidenced by stone tool manufacture (Wynn, e.g. 1979; cf. Foley 1987), or burial of the dead (e.g. Belfer-Cohen and Hovers 1992; Bergounioux 1958; Peyrony 1921; Smirnov 1989; Vandermeersch 1976). These are all very interesting avenues of investigation, but all are flawed to a greater or lesser degree by implicit assumptions about past hominid behavior, based on present understanding (cf. Davidson and Noble 1993; Gargett 1989a, 1989b).

What all this means for the present research can be summarized as follows. Dibble's, Rolland's, and Davidson and Noble's work leads to questioning of the cognitive abilities of pre-modern hominids. They lead one to consider the possibility that one may be dealing with a very different kind of cognition in pre-modern hominids. If in fact pre-modern hominid cognition was different from that of modern humans, then models used to interpret pre-modern hominid cognitive abilities need to be strongly warranted. Dibble's and Davidson and Noble's new methods are a step toward resolving the stalemate over when modern humans first arose. If it is ever to resolve the debate over the emergence of humanness, the archaeological community needs more Dibbles and Rollands and Davidsons and Nobles. The quagmire of antithetical construals of pre-modern hominid archaeological materials persuades me that the discipline needs more ways of "getting at" the cognitive abilities of pre-modern hominids far more than it needs additional fossils and artifacts as Klein (1992), for example, has stated.

Summary and Preview

In this chapter, I have suggested that, at present, archaeologists work from a limited range of models in framing their interpretations of spatial patterning in pre-modern hominid contexts. These models may ignore the possibility that one may not be dealing with modern human cognitive processes among pre-modern hominids. Current models for pre-modern hominid use of space are constructed, quite naturally, by analogy to experience with fully modern humans. These models focus on human abilities, yet little is known about the behavior of non-hominids with respect to space. With no way of knowing whether or not models based on experience with fully modern humans are useful for interpreting pre-modern hominid spatial patterns, and little in the way of information about the potential for non-hominids to create spatial patterns, archaeologists may not be able to say with any certainty that the spatial distributions of artifacts in Middle Paleolithic hominid sites are telling them anything at all about the cognitive abilities—whether modern or otherwise—of those hominids.

For the most part, spatial analyses of Middle Paleolithic localities have yielded ambiguous, or unenlightening results. This may be because spatial analysis is an inherently flawed method of determining behavior, or because the behavior of Middle Paleolithic hominids is so little like that of modern humans that their archaeological record is not structured in the same way and thus cannot provide comparable data. This is something that awaits empirical investigation. I hope this book contributes in some way to progress in that direction.

Documenting the ability of other animals to create spatial patterning could have profound implications for interpretations of spatial patterning in pre-modern hominid archaeological sites. For, if non-hominids can leave spatially patterned traces, the models used to interpret pre-modern hominid sites will have to be amended. Moreover, in order to use spatial patterns to infer pre-modern hominid cognitive abilities, meaningful differences (if any exist) will have to be sought between the patterns found in non-hominid contexts and those in caves occupied by hominids.

My research fits within the context of all such work aimed at discovering the processual links between the static phenomenon of the archaeological present, and the dynamic phenomena operating in the past. It seeks empirical justification for the hypothesis that acultural animals can and do create spatial patterns in caves. Finding such evidence would make it unnecessary to continue to invoke culture or cultural mediation as the reason for the presence of nonrandom spatial patterns in pre-modern hominid archaeological sites, and may instead provoke novel examinations of the processes that created them.

The rest of this book will present the results of a taphonomic appraisal of the study sample, and the outcome of the spatial analysis. Chapter 2 concerns itself with the method and theory of site formation processes, and with the prospects of using cave bears as subjects for spatial analysis. Chapter 3 involves the search for and accumulation of data appropriate to this study. Chapter 4 is the first of a three-part assessment of the processes most likely to have created the bone accumulation under examination. It deals with species composition and age structure of the faunal assemblage. Chapter 5 continues the assessment, with an examination of skeletal element frequencies. Chapter 6 completes the taphonomic analysis, with observations of bone modification. Chapter 7 presents the results of a qualitative and quantitative spatial analysis of the sample from Pod hradem Cave. The final chapter briefly summarizes the research presented here, offers some conclusions regarding the efficacy of using spatial analysis in studies of modern human origins, and some suggestions for future research.

A Note on Structure. Because of the complexity and mass of theoretical and methodological considerations underpinning the various segments of this analysis, I feel it would be counterproductive to present a single treatment of all the issues, followed by the results, and concluded with a monolithic discussion of those results. I have chosen instead to structure each chapter to include a presentation of the method and theory pertinent to that chapter, followed by a treatment of the empirical observations germane to the segment. Then I present the results, which, depending on the circumstances, will either include a running interpretation, or a unitary section discussing the implications of the results.

Notes

1. The date of 40 kyr ago for the beginning of the Upper Paleolithic is by no means firm, although behaviorally modern humans are in evidence by this date in Africa, Europe, Asia, and Australia. What it means to be behaviorally modern, or whether such terms as "Upper Paleolithic" are reified categories, are the crux of scholarly disagreement on the question of modern human origins. The point of introducing the traditional rubicon of the Middle to Upper Paleolithic transition here is to indicate that (most would agree) since ca. 40 kyr ago modern humans have been the sole exponents of the hominid lineage.

CHAPTER 2

Site Formation Processes and Cave Bears

Archaeologists do not "dig up" the past. Instead, they encounter and must interpret the material and contextual traces of a host of processes, both archaeological and natural, that comprise the archaeological universe. Binford and Binford (1968) recognized that archaeology seeks knowledge of the past through what can be observed in the present. This insight prompted research into "site formation processes" and "taphonomy," the work of coming to know the causal links between dynamic processes occurring in the past, including human and hominid behavior, and the static, present-day phenomena of the archaeological entities we study. Thus, for the last 20 years much archaeological research has been concerned with what Binford (1977, 1981b) has called the middle range of theory building. This kind of research has expanded the range of sources for the analogues and models that are the foundation of archaeological interpretation, through observation of modern forager societies, extant animal populations, and the chemical and geological processes that can affect the preservation of archaeological materials (e.g. Behrensmeyer and Hill 1980 and papers therein; Binford 1977, 1978a, 1978b, 1981b; Bonnichsen and Sorg 1989 and papers therein; Brain 1981; Gifford 1977; 1981; Schiffer 1972, 1987; Yellen 1977). Such studies provide the wherewithal for archaeologists to make more secure inferences about the past.

Inferring the causes of patterned archaeological traces based on knowledge of the present is an example of the uniformitarian method first seen in the works of Hutton and Lyell in the 18th and 19th centuries (see Simpson [1970] for a treatment of the history and methodology). Uniformitarian reasoning presumes that processes making traces of a certain kind in the present most likely left similar traces in the past. However, given that different processes can produce similar effects (a phenomenon known as equifinality), understanding the range of potential contributors to site formation is crucial, because archaeologists need the ability to identify processes capable of masking or mimicking patterns created by humans or hominids. In this regard, natural, or non-hominid-related processes in the present have received much attention, because it is axiomatic that archaeologists must first rule out other causes for the traces and patterns they observe before attributing them to humans or hominids.

Employing a uniformitarian framework, the research presented in this book is aimed at further specifying the range of processes that can produce spatial patterning in pre-modern hominid archaeological sites. At present, not enough is known about the material consequences of cave-dwelling animals' activities to say that all spatial distributions at a hominid cave site are necessarily the product of cultural abilities. This work explores a cave bear hibernation den, with an eye toward discovering whether or not those bears were capable of creating spatial patterns where they hibernated.

This chapter lays the foundation for my taphonomic analysis of cave bear fossils from Pod hradem Cave. First, a brief overview of the study of site formation processes, especially those in caves. This will ground the present research in the efforts by archaeology and other historical disciplines to infer processual causes for the observations archaeologists make. The second part of the chapter considers the suitability of using cave bears for such a study.

Site Formation Processes

Research into site formation processes has brought about profound changes in archaeological theory and practice. Prior to the 1970s, archaeological interpretation employed common-sense assumptions, often implicit, about the processes responsible for creating and structuring the archaeological universe (see e.g. Brew 1946:45; Kluckhohn 1939:331). With the coming of the New Archaeology in the 1960s and 70s, archaeologists began to seek *explicit* links between the structure of an archaeological site and the processes that created it.

This program arose when it was recognized that, although the past is the subject, the objects and relationships archaeologists analyze are contemporary phenomena (Binford and Binford 1968). The epistemic question is thus, how does one move from observations on the static present phenomenon of archaeological materials and relationships to statements about past dynamics? The answer lies in developing middle-range theory regarding the formation processes of the archaeological record (Binford 1977:10). This insight was taken up by Schiffer (e.g. 1972, 1976, 1983, 1987), as he developed the concept of site formation processes.

In middle-range theory building, structured similarities are first sought between the archaeological subject, which is a product of past processes, and present-day processes that give rise to products similar to those under investigation. The second stage of middle-range reasoning involves the inference that processes similar to those observable present-day processes were likely to have been operating in the past, thus giving rise to the archaeological traces. Where human behavioral processes are concerned, i.e. those which may produce archaeological traces, ethnographic observation of behavioral regularities among modern humans may be used to explain the particulars of archaeological observations. And, since many archaeological sites have been formed as much by natural processes as by human or hominid actions, research into present-day non-cultural processes can contribute much to the theory of individual site formation.

In North America, there have been three major streams in this research program. In Schiffer's work, "site formation processes," as he defined them, figured most prominently (Schiffer 1972, 1976, 1983, 1987). His theoretical treatises formalized some of the links between cultural and natural processes and the structure of archaeological sites. In another, closely related, domain Binford's (1977, 1981a, 1981b) ethnoarchaeological work with the Nunamiut of Alaska was explicitly formulated to elucidate the patterning in European Middle Paleolithic archaeological sites, by investigating the variability present in the archaeological remains of sites having different functions among the Nunamiut. Finally, the paleontological subfield of taphonomy has, in the hands of archaeofaunal specialists; provided a great deal of useful information about the processes that intervene between the time an animal dies (either from natural causes or predation) and the moment its remains are recovered by the archaeologist (e.g. Behrensmeyer 1973; Behrensmeyer and Kidwell 1985; Brain 1981; Gifford 1977, 1981).

All of these enterprises seek similar kinds of knowledge, but the paths to understanding are diverse. Some, like Blumenschine (1986, 1987, 1988, 1989), Haynes (1980, 1983a, 1983b, 1985, 1988a, 1988b), Hill (1976, 1977, 1979, 1980, 1983, 1989) and Binford (1981b), have studied contemporary carnivores to compile diagnostic criteria of the ways in which they modify bones. This knowledge is then used to help distinguish the actors and agents that created ambiguous bone accumulations in prehistoric localities. Others, like Behrensmeyer (1975, 1978), Lyman (1984, 1985), and Voorhies (1969), have concentrated on the physical properties of bone and the geological forces that can act to degrade and disturb bones in the environment, in order to better understand the likelihood of preservation under various conditions.

Research into site formation processes, in essence the "taphonomy" of archaeological sites (Gifford 1981), is indeed crucial to the health of the archaeological enterprise. It enables archaeologists to make more secure inferences about the past from the materials they study. Although there is no such thing as perfect knowledge of the past, knowledge based on systemic links between causal agents and observable physical manifestations is probably more reliable than knowledge constructed in the absence of such an understanding (e.g. see Gifford-Gonzalez 1989; Wylie 1982, 1985, 1989a). The research reported here was undertaken from this same perspective. I wish to bolster the warranting arguments for interpretations of spatial patterning in pre-modern hominid sites. To do this I am further specifying the range of sources for analogies with which to interpret archaeological patterning. As explained in Chapter 1, present interpretations of spatial patterning depend to a large extent on models of how modern humans behave in space. The results of this investigation of an extinct non-hominid's bones may help clarify the effects of its behavior. Moreover it will provide a base-line, of sorts, against which archaeologists can compare the patterning observed in pre-modern hominid sites. By demonstrating that non-cultural behaviors can produce patterns similar to those observed in pre-modern hominid sites, knowing that cave bears produced non-random horizontal spatial patterns in their hibernation dens will contribute to our understanding of the determinants of spatial structure in pre-modern hominid sites.

Why Caves?

There are two reasons for focussing on caves and spatial patterning in caves. First, like other mammals, pre-modern hominids sought shelter in caves. By reference to knowledge of extant animal species, it is possible to speculate on the reasons an animal might seek shelter in caves beyond those that might be learned from conversations with modern humans. Caves are used as safe places by many animals, including many carnivores, and for a variety of reasons, among them: giving birth and raising young; protection from predators during sleep; places of refuge during consumption of prey, for protection from intraspecific competitors. It is highly likely that, regardless of the cultural abilities of pre-modern hominids, all of the reasons given above for animal use of caves would also be logical reasons for the hominids to seek shelter in caves. Second, caves are important if only because so much of what we know or claim to know about hominid behavior during the later Pleistocene of Europe comes from cave deposits. For these reasons, caves are appropriate contexts from which to draw for analogical sources.

Cave Taphonomy

Most discoveries of European Middle Paleolithic hominid fossils and artifact assemblages have been made in caves and rockshelters. Any effort, therefore, to understand the formation of sites within them must include a consideration of the processes affecting these natural shelters before, during, and after occupation, and which render them especially good sedimentary environments for bone preservation. A substantial amount of literature regarding the geomorphology of caves exists (e.g. Dreybrodt 1988; Ford and Cullingford 1976; Ford and Williams 1989; White 1988), and a great deal has been written about archaeology in caves (e.g. Butzer 1981; Collcutt 1979; Farrand 1985; Laville et al. 1980; Straus 1979, 1985, 1990; Villa 1982; Villa and Courtin 1983). The following is a very brief overview.

Although caves and rockshelters may be formed in sandstone or other dry sedimentary rocks, the vast majority of caves containing traces of European Middle Paleolithic hominids are formed in limestone and similar calcareous sedimentary rock. Such caves usually form by the gradual deletion, through solution and mechanical processes, of sediments of various sizes from the bedrock. Many of the same formation processes apply equally to caves and rockshelters, and for the sake of brevity, I shall refer hereafter only to caves. Caves should be viewed as a unique kind of site, because particle size and origin are potentially more variable in caves than in any other sedimentary environment (Jennings 1985:163). Also, no two caves are formed in identical parent material, nor under identical circumstances (Farrand 1985:23). Each thus presents a novel interpretive problem in the reconstruction of its depositional history.

Differential susceptibility to weathering will result in very different depositional histories for two caves of otherwise similar morphology. For example, massive, horizontally bedded limestone will tend to break down more slowly than thinly bedded and steeply dipping material. The more thinly bedded the bedrock, the more gradual the breakdown, and the smaller the particles produced. On the other hand, more massive material will tend to break down by a process of periodic, sometimes catastrophic, collapse. At these times, the sediments produced can range from dust to house-sized blocks (Ford 1976).

Caves may house deeply stratified deposits, and a great many contain material of paleontological or archaeological interest, or both. Cave deposits can build up in many ways. Gradual accumulation of sediments can occur on the floor of the cave as the walls and ceiling gradually break down due to chemical weathering, frost, seismic disturbance, and so on (Ford 1976:51). The rate of such deposition is highly variable. Some caves, for example Shanidar (Solecki 1955), contain many meters of horizontally stratified deposits, while others, for example La Chapelle-aux-Saints (Bouyssonie et al. 1908), contain much thinner deposits.

Sediments from outside can be introduced, either through chimneys to the surface above, or through the action of running water or wind, carrying allochthonous sediments into the cave. Wind can introduce various kinds of sediments, the most common of which is loess. Pollen and fragile plant macrofossils are also found in caves, and when conditions are optimal, many normally open-air plant species can live in the sunlit portions of caves, not only contributing their remains to the deposits, but also potentially disturbing previously deposited materials with their roots. Most allochthonous sediments are recognizable as such through application of geological techniques, including particle size analysis, bedding, and petrographic analysis. Knowledge of the origins of sediments in caves is crucial to an understanding of the archaeological materials contained in them.

All of the physical and erosional mechanisms that can act on open-air site sediments can also work in caves, such as downslope movement due to gravity, surface water or ice (Butzer 1982:54).

Gullying by runoff, and subsidence in sinkholes due to groundwater, can also occur (Farrand 1985:25; Glover 1979). Hominids and other animals moving about in the cave, especially in a moist, soft substrate, can vertically displace objects by as much as 10 cm (Butzer 1981:155; Gifford-Gonzalez et al. 1985; Villa and Courtin 1983). Even buried remains may not escape disturbance (Harrold 1980:196; Trinkaus 1985:209). It is even possible for stratified cave sediments to be mixed completely (Erlandson 1984; Wood and Johnson 1978). Once hominids have left a cave, subsequent animal inhabitants are capable of further disturbing the integrity of the archaeological deposits.

Because they may be used by numerous animal species, as well as by hominids, caves can accumulate bone rapidly (Farrand 1985:22; Henschel et al. 1979; Klein 1975:286; Straus 1979:333; Sutcliffe 1970:1110; Sutcliffe et al. 1976:495). Animals dying in caves, and those carried in and consumed by denning species, are equally likely to have become part of the cave deposits. Separating hominid from other animal agents of accumulation is therefore a major archaeological task.

Caves often contain more extensive and better preserved bone accumulations than open-air locations. There are several reasons for this. First, relatively rapid deposition is common in caves, much more so than in the open. Once a specimen is buried, the likelihood of its being preserved is enhanced. Second, a karstic cave's protected environment, along with its somewhat alkaline chemistry, contribute to a greater likelihood of bone preservation. In addition, Kurtén (1976:88) has noted that bones can find their way into exceptional, protected locations, such as natural niches or against walls, within caves. Whether they arrived there by chance, or through "housekeeping" on the part of cave inhabitants, they stand a better chance of survival than bones lying out on the cave floor, where they are susceptible to fragmentation due to trampling. Finally, nonuniform processes within any cave can contribute to the creation of micro-depositional environments, such as natural depressions in the floor, where sedimentation can be more rapid than elsewhere. Such places will also be more likely to preserve bone compared with specimens left exposed on the cave floor.

In sum, caves are locales with inherently complex natural depositional histories. Cave deposits are often the result of multiple agents of deposition. They thus present difficult interpretive problems for any archaeologist wishing to study their contents.

Animals in Space

Kerbis-Peterhans (1990) and Lam (1992) have reported the presence of patterning in the horizontal distribution of bones within hyena dens. Although their studies were not undertaken with the same research interests as the present study, they nevertheless reveal behaviorally meaningful spatial patterning in the bony remains at hyena dens. To my knowledge, these are the only two studies documenting spatially patterned remains in caves occupied by animals. And whereas Kerbis-Peterhans and Lam's work represents evidence of the ability of animals to produce clustered spatial patterns in the materials brought to their dens, data collected and analyzed systematically with the explicit research goals I have outlined will usefully expand understanding of the phenomenon of spatial patterning in animal dens, both by including another species—cave bear—and through the use of a much larger sample, both in terms of the number of specimens and the horizontal extent of the distributions.

Cave Bears

Ursus spelaeus, the cave bear, was a sister species of the extant brown bear (*Ursus arctos*) (Kurtén 1976). Their fossils are found across Europe, from northern Spain to the eastern Caucasus, and from southern England to southern Italy and Greece. They are first seen in the fossil record about 300,000 years ago, during the Holsteinian Interglacial. Thus, the species persisted through several climatic shifts. Nevertheless, they were all but extinct by the time of the terminal Pleistocene deglaciation. During the most recent cold period, they were associated with a cold fauna, including wooly mammoth, wooly rhinoceros, musk oxen, reindeer, cave hyena, wolf, wolverine, arctic fox, and cave lion.

There is no good evidence that cave bears were hunted by hominids (see e.g. Koby 1953; Kurtén 1976; but see Gábori-Csánk [1968], on the Middle Paleolithic site of Erd, in Hungary, for a contradictory opinion). Although they may have had no direct effect on their populations through predation, Kurtén (1958) postulates that modern humans, who seem to have enjoyed the shelter of caves, gradually eliminated caves as secure and perennial hibernation locales of the bears, leading to increasing isolation of viable populations and eventual extinction.

Although they were large mammals, known to inhabit caves, and to leave their bones in huge numbers, the extinct cave bear has received little attention from archaeologists attempting to enhance knowledge of site formation processes and bone modification. This is most likcly because bears are not known to accumulate bone in their occupations, nor are their bone modification abilities great. Besides their living in caves, I chose to explore cave bears for a number of reasons: their body size, behavior, the availability of large collections, and most importantly their obvious lack of culture as it was described in the introduction.

Body Size

Hominids seeking shelter in a cave would have been constrained by its walls and ceiling. In other words, their movement would have been restricted to those portions of the cave that afforded sufficient headroom and floor space to carry out activities. Hominids would also have sought areas with little or no slope on which to rest, and sufficient light for their activities. Such constraints may be important determinants of the spatial patterning of archaeological sites. Indeed, it was a concern with the physical parameters of human movement and limitations on their actions that underpinned Binford's (e.g. 1978b, 1983: chapter 7) contributions to the study of past human and hominid space use. Thus, to ensure that the present research was useful as a source for models used to interpret pre-modern hominid space use, the subject of my study had to be an animal with similar space requirements as the hominids whose behavior I ultimately wished to understand better. Indeed, the size of *Ursus spelaeus* makes it a suitable subject for this research.

Cave bears were the largest ursids that ever lived. Kurtén (1976:25) estimates that a large adult male might have weighed 450 kg in a lean state, and much more prior to hibernation. However, despite the extreme disparity in body mass between hominids and these ursids, the cave bears' height when on all fours, and length when curled up to hibernate would have been closer to those parameters in recent hominids than any other large mammal inhabiting caves. Cave bears would thus have required about the same diameter of space on the ground to hibernate, and they would have needed nearly the same head-room during ingress and egress as recent hominids. Whatever "creature comforts" the bears required would thus have been closest of all other cave-dwelling mammals to those of late Pleistocene hominids, whether pre-modern or modern.

Hibernation, Caves, and Site Formation

In caves throughout the karst regions of Europe cave bear bones are often found in numbers far exceeding those of other large mammal cave dwellers. Winter mortality and favorable preservation conditions in caves combined to bring about this situation. It is therefore useful to review how ursid behavioral adaptations interact with geomorphology to produce the peculiar characteristics of a cave bear assemblage.

Like many animals in seasonal ecosystems, bears are sensitive to winter shortages of food (Kurtén 1976; Rogers 1987), so much so that Ursidae have evolved the unique adaptation of hibernating. Observations of modern wild brown bears enable us to model the extinct cave bear's behavior. Extant brown bears are capable of lowering their metabolic rates almost to zero, conserving body stores of fat and other nutrients when forage is least plentiful. Mortality studies reveal that they are at their most vulnerable during dormancy and immediately following (Rogers 1987). Old age, debility and disease, coupled with annual fluctuations in food availability leading to poor foraging prior to dormancy, combine to cause death during hibernation. Among living species, cubs are particularly vulnerable. Newborns do not hibernate, and are therefore in need of constant high levels of nutrition. If their mother has insufficient stores of fat, and is unable to sustain lactation, cubs die. It can be reasonably assumed that cave bears also endured the Pleistocene winter using the hibernation strategy, and that patterns of behavior and mortality similar to those of extant brown bears would have been the case for *Ursus spelaeus.*

Knowing the foraging habits and behavior of modern brown bears also helps to further refine understanding of the conditions under which cave bear bones came to be preserved in large numbers during the Pleistocene. Modern bears hibernate in a variety of protected spaces. Most hibernate in hollow trees, under fallen trees, in earth overhangs, or in snow dens. Some, in areas where competitors and humans are rare, still find small caves in which to overwinter. Although cave bears very likely employed a similar range of hibernation places, and may have hibernated in caves as only one of a number of alternatives, their species is known only from well-preserved assemblages from the karstic caves of Europe (although the cave bear fauna from Erd, Hungary, may be the sole excavated exception [Gábori-Csánk 1968]).

During the hibernation period, which can last up to six months, extant bears do not usually leave the hibernation locale. Given a longer cold period, winter mortality would likely have affected cave bears more often than not *while still in the cave*. Preservation conditions in the karstic caves ensured that when a cave bear died during hibernation, its bones were highly likely to be preserved. Thus, as a result of the interaction of winter mortality and good preservation, cave bear localities are legendary for the numbers of individuals unearthed from deposits in the caves they occupied, throughout the Upper Pleistocene. Indeed, the number of specimens in even a small collection from a single site can number in the tens of thousands, representing hundreds of individuals (e.g. Pod hradem Cave, the subject of this investigation). The larger cave bear localities boast tens of thousands of *individuals*. For example, the Drachenhöhle near Mixnitz, Austria, is estimated to have contained the remains of 30-50,000 individuals (Abel and Kyrle 1931)! The sheer numbers of cave bear specimens available for study from many localities thus promise samples sufficiently large to yield robust statistical inferences about the presence of any spatial patterning, with the potential to provide answers to the puzzle of which agent or agents created those patterns.

Bears and Bones: Accumulation, Modification, and Space

Suitability of cave bears for this research is enhanced by what can be inferred about their behavior with respect to bone. To best interpret the faunal assemblage characteristics, it is important to understand whether or not cave bears might have accumulated the bones of other species in their hibernation locales, whether and to what extent they were capable of modifying the bones of deceased conspecifics or others that had entered the cave through whatever means, and how the behavior of the bears themselves might have affected the spatial distribution of all bones in the cave. In all cases, the verdict on cave bear behavior is favorable to this investigation.

One important aspect of cave bear behavior making the species interesting for this examination is that evidence strongly suggests they did not collect other animals' bones in their cave refuges. This contrasts with habits of other cave-frequenting carnivores such as wolves and hyenas. Although it cannot be simply assumed that cave bear behavior is analogous to that of modern bears, at least two independent lines of reasoning imply that they did not collect bones in their dens. The first is analogy to modern bear behavior, and the second is from studies of the diet of cave bears, carried out through isotopic analysis of sub-fossil cave bear specimens. Together, these produce a picture of a species that was even less likely than extant ursids to collect bones.

First, there has been no documentation of modern hibernation dens containing bones, nor of bears carrying bones back to their dens for consumption. Even the more carnivorous species of ursid, the brown bears (including the Grizzly), are not known to introduce bones into their hibernation dens. Second, cave bear diet is thought to have been almost exclusively herbivorous (Kurtén 1976; Bocherens et al. 1990). Maxillofacial architecture, dental morphology, and dental wear all imply that cave bears were specialized plant eaters (Kurtén 1976). Moreover, recent isotopic bone chemistry analyses have lent support to this conclusion. Bocherens et al. (1990) performed $\delta^{13}C$ and $\delta^{15}C$ analyses of sub-fossil cave bear bones from Aldène Cave, in France. Their study places the cave bears in a group with a number of obligate herbivores such as *Bos*, *Ovis*, and *Rupicapra*, and distinct from meat-eaters such as *Lynx*, *Canis*, and *Lutra*. It seems likely therefore that bears consuming vegetation to the same degree as cows and sheep would not have collected the bones of other animals, especially if the more omnivorous extant bears do not behave in this way. Thus, on the basis of skeletal morphology, dental attrition, bone chemistry, and analogy to modern bear behavior, it seems that cave bears probably did not introduce the bones of other animals into their caves. It is most likely, therefore, that cave bears contributed only their own skeletons to the assemblages of bones that accumulated in their dens.

There is the possibility that cave bears gnawed bones in their sites. Haynes (1983b), for example, has observed that modern bears are capable of their own characteristic form of bone modification, through gnawing with their molar teeth. Predominantly herbivorous *U. spelaeus*, however, might have gnawed bone only rarely, if ever, in contrast to the omnivorous extant species that Haynes studied. This possibility will be investigated in Chapter 6.

It is most likely that the cave bear's greatest impact on previously deposited (and perhaps buried) bones would have been in its ability and propensity to displace them during preparation of a place to hibernate. Rogers (1981) observed modern black bears preparing nests of vegetation in burrows and caves in order to make a dry, comfortable space to rest for up to six months. In one example, a female and her two cubs worked for some time, first dragging all the detritus out of the space, then moving in the new vegetation, then dragging it all out again before producing a "satisfactory" bed (Rogers 1981:68). Thus, although modern bears are not potential bone consumers, like hyenas and wolves, by analogy to the modern ursids, the propensity for cave bears to spatially rearrange (and incidentally modify) bones would no doubt have been high.

Trampling might have had a profound effect on the bones already in the cave sediments. This would not have been the case for rodents or other smaller mammals living in caves, nor for hyenas or wolves, all of whose much lighter bodies would not have had as great an effect on most size classes of bone in their dens. However, cave bears, weighing more than 500 kg would very likely have put great stress on most bones in their way. Bahn (1983), for example, suggests that trampling by bears was responsible for the creation of so-called bone buttons, frequently found in cave bear accumulations. These are fragments of bear fibulae that appear to have been stepped on in soft substrates, and, due to the structure of the fibula, all fractured in similar ways, leading early workers to hypothesize that they had been manufactured by hominids. Bahn argues cogently for a natural process creating these objects. In another example, Trinkaus (1985:209) suggests that it may have been cave bears that fragmented the Neandertal skeletons at Krapina, although this conclusion must await a thorough taphonomic assessment.

An Acultural Model

Finally, because cave bears did not have a culture, at least as defined in Chapter 1, it is therefore possible to assess behaviorally generated spatial patterning in the absence of culture. It should be safe to infer that at least some of the patterning discovered in cave bear sites can be attributed to the ursids' instinctual and learned behaviors, even if some carnivore or even hominid disturbances are possible. Knowing something about spatial patterning created by non-cultural behavior in a species coeval with pre-modern hominids is very relevant to Middle Paleolithic studies.

Summary

Study of site formation processes is an integral part of archaeological practice. This chapter has outlined the methodological underpinning of the present research—i.e. uniformitarianism—and stressed the importance of understanding the theoretical links between knowledge of present-day processes and those inferred to have been at work in past contexts. The more known about the range of processes capable of depositing, redepositing, modifying, or preserving archaeological materials, the better archaeologists will be at making strong inferences about the processes at work in the past that brought about the material and contextual traces they study in the present.

The choices of study context and animal subject were also justified in this chapter. Caves are unique sedimentary environments, and can contain large numbers of well-preserved bone specimens of paleontological and archaeological interest, because both hominids and non-hominids alike shelter in them. They have also been the source of much of what is known about the Middle and Early Upper Paleolithic of Europe and elsewhere. Finally, to begin to address the issue of whether or not the spatial patterning found in Mousterian hominid cave sites is the result of culture as the term is used in this book, a baseline for comparison is needed. Cave bears are approximately the right size, apparently behaved in such a way that facilitates study of their space use, died in sufficiently large numbers in caves to permit spatial analysis, and did not have a cultural capacity. The presence of large numbers of cave bear fossils in the karstic cave of Pod hradem, Czech Republic, made it a candidate for this study. The next chapter introduces the site and its qualifications.

CHAPTER 3

The Search for Appropriate Data

This chapter begins with a consideration of the criteria employed in the search for appropriate data—excavation strategy, recovery tactics, and so on. Next, the reader is introduced to Pod hradem Cave, with a brief account of its excavation history and the depositional, archaeological and paleontological findings. Then I give my rationale for choosing the sub-sample on which I performed the analysis.

Site Search

Because this work is, at its heart, a study of the degree to which cave bears spatially structured the skeletal remains in their caves, I needed to rule out any possibility that culturally mediated behaviors on the part of hominids had contributed to the spatial distributions I was monitoring. Several criteria had to be met by any collection before it would be of value in such a study. First, I needed to find a collection from a cave with no evidence of human or hominid activity. Thus, the stratigraphic unit from which the study sample originated could have no artifacts or features in it. Preferably, the unit chosen would have at least one nonartifactual level between it and any artifact-bearing strata.

Although the absence of stone tools or features cannot be used as evidence that hominids never visited the cave during deposition of the strata of interest, the absence of such evidence nevertheless makes it difficult to argue that their activities heavily influenced the spatial distribution of materials deposited during that time (but see Brugal and Jaubert 1991 for a discussion of low-yield archaeological sites and the possibility of systematic hominid scavenging). Because so many European caves contain at least some levels with evidence of hominid presence, and many with hominid and other animal occupations overlapping, this first criterion severely restricted the search. Unlike Stiner (1990a), who employed convergent lines of evidence to determine the *predominant* agent of accumulation where both hominids and animals contributed to site formation, I sought an unequivocally non-hominid site.

Secondly, statistical techniques designed to discover or verify any spatial patterning require both large numbers of individual specimens, and sufficient numbers of provenience units on which to base their calculations. Thus, I required material collected from a substantial horizontal excavation within a cave. Moreover, in order to monitor any parameters that may have been determined by proximity to walls, such as specimen size, I reasoned that the sample I sought had to encompass at least half of the area circumscribed by the cave walls. In addition, the excavation must have extended all the way to the walls, because of the possibility of finding more bones of all sizes nearer the wall than would be contained in sediments away from walls. That this criterion was perhaps easiest to satisfy is a mixed blessing. A great many Pleistocene cave localities in Europe have been mined in their totality, which meant that on this criterion alone I would have had many candidates. It is, however, to the great credit of the excavators of Pod hradem Cave that they intentionally left unexcavated one half of the cave deposits.

Third, in order to apply even rudimentary statistical tests for clustering, I also required material collected with control over horizontal provenience. The minimum necessary was a 1 by 1 meter grid and some form of level control, either arbitrary or natural. Archaeological excavations in Europe since the 1960s have generally employed three-dimensional point provenience recording techniques, but many paleontological excavations until very recently took few pains to record all but minimal provenience data. Since what I sought was, essentially, the archaeological investigation of a paleontological locality, my search was made vastly more difficult.

In the end, after having canvassed the leading archaeologists and paleontologists of Europe, only one site to which I could have had access offered the minimal hominid presence, maximal horizontal extent of the excavation, and data recovery standards that I had set out to find: Pod hradem Cave.

Pod Hradem Cave

Pod hradem Cave lies approximately 25 km NNE of Brno (49°22'22" N; 16°43'23" E), in the Moravian karst region (*Moravský kras*) of the Czech Republic (Figure 3.1). The cave is one of several important paleontological and archaeological localities in the area of the source of the Punkva River, which emerges from its subterranean course nearby—most notably Kůlna (Valoch et al. 1969). Pod hradem's entrance is to be found in the north valley wall, beneath the ruined Castle Blansek (Pod hradem means "under the castle" in Czech).

Pod hradem is a horizontal solution tube, developed in Devonian limestone, that became open to the air sometime in the Middle Pleistocene (Musil 1965). The main tunnel's diameter varies from a minimum of about 3 to about 4 to 5 meters wide at the level of the historic cave floor. A wider chamber, about 30 meters from the entrance, provided the opportunity to open an extensive excavation, making the collection suitable for this research (see Figure 3.2).

Excavation History

Although amateur fossil hunters had plumbed the cave for its contents in several places in the 19th century, the first systematic excavations were undertaken in 1890 by Knies (Knies 1901, cited in Musil 1965). Knies was later joined by Trampler (1898, 1900, cited in Musil 1965). They noted that much of the present-day sediments in the cave derived from the chimney that opened to the plateau above. Knies reports the preponderance of *Ursus spelaeus.* Species identified in Knies and Trampler's excavations were the following (those noted with an asterisk were also recovered in the modern excavations by the Moravské Museum): *Falco perigrinus*, *Cerchneis tinnunculus*, *Nystea scandiaca*, *Nucifraga caryoçatactes*, *Scolopax rusticola*, *Vanellus vanellus*, *Talpa curopaea*, **Lepus* sp., *Lynx lynx*, **Panthera spelaea*, **Crocuta crocuta*, **Canis lupus*, **Martes* sp., **Ursus spelaeus*, **Equus* sp., **Rangifer tarandus*, **Bison priscus*, **Bos primigenius*, **Rupicapra rupicapra*, **Capra ibex.*

The modern excavation of Pod hradem, which yielded the collection examined in this study, occurred during the summers of 1956, 1957, and 1958. Drs. Rudolf Musil and Karel Valoch, then of

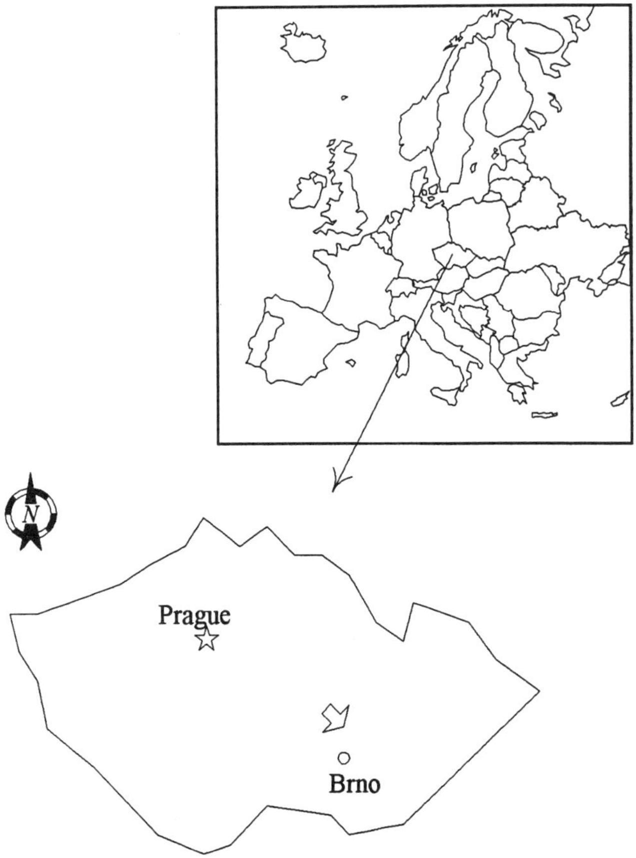

Figure 3.1. Pod hradem Cave: location. Arrow points to site.

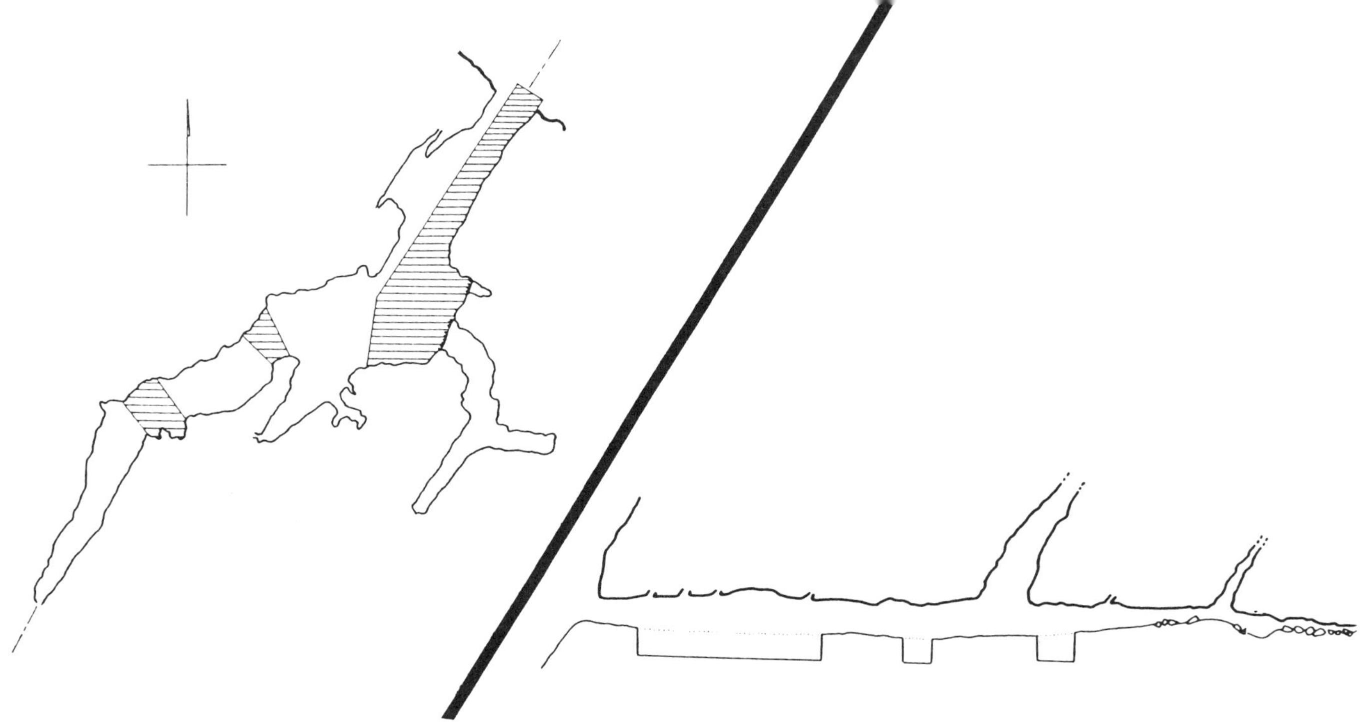

Figure 3.2. Pod hradem Cave: ground plan and cross section. Hatched areas are approximate locations of the Moravské Museum excavations, 1956-1958 (after Panoš 1965:142-143). Scale: 1 cm = 6 m.

the Mährischen Museum (later the Moravské Muzeum), investigated the cave for its paleontological and archaeological potential. In those three seasons of excavation, a deep trench some 30 meters long was opened from the vestibule to the main chamber (see plan, Figure 3.3). All excavated sediments were examined for their contents, and all materials of interest were conserved according to the square meter and 10 cm level from which they were taken. All pieces of bone were kept, including undiagnostic fragments. Thus, the collection is capable of answering the questions I am asking. Considering that this material was retrieved nearly forty years ago, the quality of data is remarkable. Those species found only during the modern excavations were: *Dafila acuta*, *Lyrurus tetrix*, *Lagopus lagopus*, *Lagopus mutus*, *Capella media*, *Strix aluco*, *Garrulus glandarius*, *Citellus citellus*, *Cricitelus* sp., *Clethrionomys glareolus*, *Microtus arvalis-agrestis*, *M. gregalis*, *M. oeconomus*, *M. nivalis*, *Arvicola terrestris*, *Vulpes vulpes*, *Alopex lagopus*, *Putorius* cf. *putorius*, *Ursus arctos*, *Mammuthus primigenius*, *Coelodonta antiquitatis*, *Sus scrofa*, *?Saiga tatarica*.

Approximately 30,000 specimens were recovered during the Moravské Museum excavations in the 1950s. Each specimen had been cleaned and varnished during curation, to preserve the bone. Although this had the effect of rendering the specimens unsuitable for surface casting for microscopic examination of modification, it is still possible to tell the difference between scratches and tooth dents, although it may never be possible to determine if the less profound scratches are merely the result of trampling or sedimentary pressures, or if they are indeed the result of chewing. I will deal with this problem in Chapter 6.

Depositional History of the Cave

As Figure 3.4 illustrates, in almost seven meters of deposits the excavators recognized numerous discrete stratigraphic units, which were discerned using criteria of sediment color and texture. Pelišek's (1965) interpretation of the sediments does not explicitly accord with the excavators' definition of strata. However, from the sedimentary descriptions he provides, it is possible to infer that his observations of depositional units correlate with those of the excavators.

Choosing a Sample

Although there are some 30,000 specimens in the Pod hradem Cave faunal collection, I could feasibly examine only a sample. The criteria for choosing the sample were as follows. First, at a minimum I

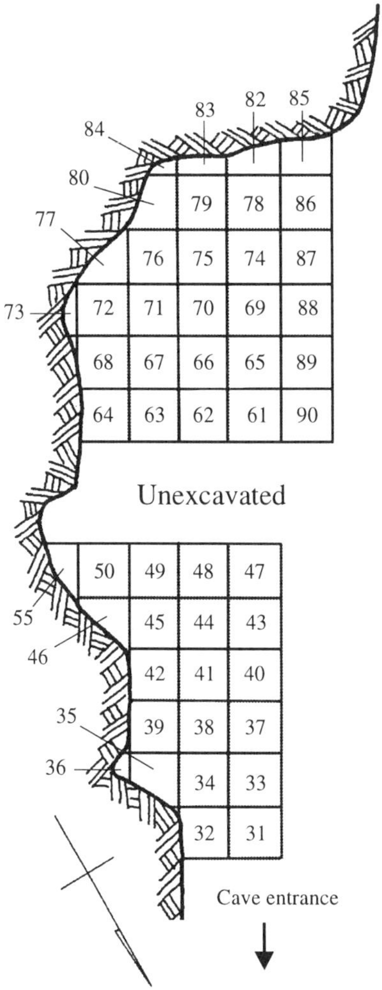

Figure 3.3. Pod hradem Cave: plan of excavation. 1 x 1 m excavation units indicated (after Musil 1965:13).

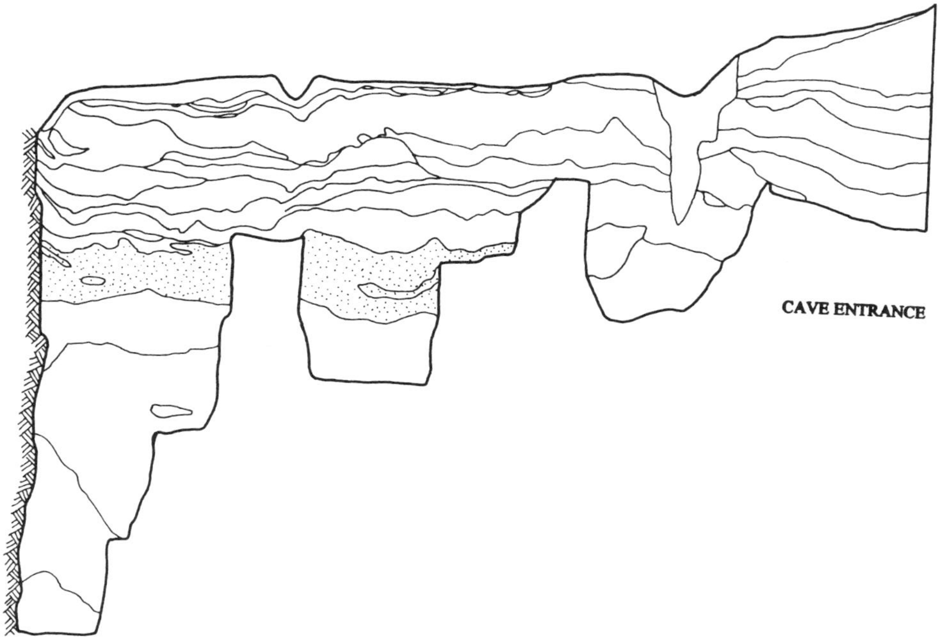

Figure 3.4. Pod hradem Cave: stratigraphic profile. Horizontal scale in meters. Vertical scale exaggerated (after Musil 1965:28A).

needed to investigate one depositional unit with enough specimens to produce robust statistical inferences. Second, because I wanted to be able to say with some certainty that the bears themselves had created any patterning that might be observed, I sought a depositional unit having few representatives of other species. This, I hoped, would ensure my using material that had undergone the least amount of disturbance or modification by animals other than the bears.

The way in which the excavation proceeded made the choice of depositional unit difficult. The excavators did not always confine themselves to a single meter square when excavating a particular 10 cm level, and often combined the material from two or more contiguous horizontal units. Having provenience units that are aggregates of two or more meter squares renders spatial analysis somewhat less robust, as discussed in more detail in Chapter 7. Thus, the third criterion was a stratum where excavation unit aggregation had occurred the least. Stratum 17 afforded the best combination of the search parameters (see Figure 3.4, in which Stratum 17 is stippled).

Depositional History and Chronology of Stratum 17

Pelišek (1965) examined samples from seven stratigraphic columns, obtaining data on grain size, and humic and carbonate content. A schematic diagram of a grain size analysis from one of his column samples is reproduced here as Figure 3.5. The sediments of Stratum 17 correspond to those in the diagram between about 200 and 300 cm below the surface. He describes Stratum 17 as a light ochre-brown clayey loam with predominantly fine sediments, about 35% weathered limestone in the 2-250 mm size range, and numerous bone remains. Cave breakdown, resulting in increased fractions of larger clasts, is inferred to be greatest during periods of deep cold. During Stratum 17's deposition, the cave was apparently breaking down at a slower rate than before or since, allowing the inference that it was deposited during a warmer period.

Radiometric dating of Pod hradem allows one to say that the cave bear remains from Stratum 17 are older than about 32 kyr B.P. Thus, it is likely that the sediments of Stratum 17 were deposited in the period of relative warmth corresponding to the middle of oxygen isotope stage 3, around 40 kyr B.P. The cave bears from Stratum 17 were alive, then, in interstadial conditions, just prior to the advent of modern humans in Europe.

Finally, as one might expect from sediments in which bears had hibernated for many years, those from this time period yielded high levels of organics. This can be attributed to the decomposition of large

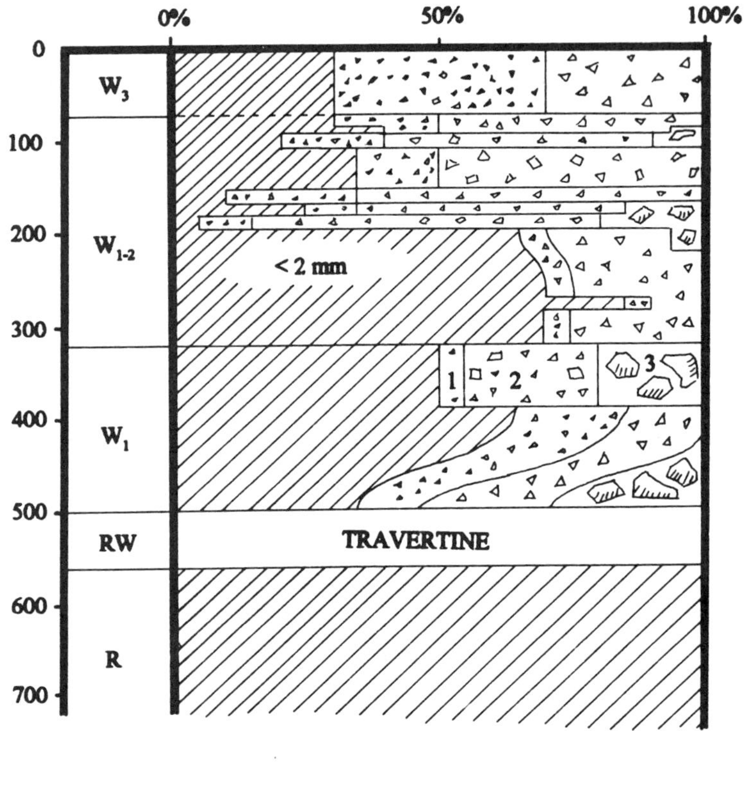

Figure 3.5. Pelišek's (1965) schematic profile of Pod hradem Cave, showing proportions of fine and coarse sediments. 1: 2-20 mm; 2: 20-250 mm; 3: > 250 mm. R = Riss glaciation; RW = Riss/Würm interglacial; W_1 = Early Würm glaciation; W_{1-2} = Würm Interstadial; W_3 = Last glaciation (after Pelišek 1965:121).

mammal carcasses, and to the urine and feces that would have built up around the hibernating animals.

Kukla (1968) concurs with Pod hradem's excavators on the climatic regime and likely chronology of Stratum 17's deposition, although he has disagreed with other chronostratigraphic inferences published in the 1965 monograph. His main contention is that the depositional complex that includes Stratum 17, which the excavators consider to have been deposited in a uniform climatic regime, does not provide evidence for this conclusion. Kukla suggests that only the very lowest layers, 17 and 18, could have been deposited in a period of relative warmth. Those above Stratum 17 exhibit evidence of cryoturbation, implying a very different climatic regime, one more in keeping with fully glacial conditions.

Archaeological Summary

Valoch (1965) reports sparse evidence for hominids in Pod hradem, which occurred only after the sediments in Stratum 17 were deposited. A very few stone tools were recovered from higher in the deposits, attesting to at least occasional visits to the cave in the Late Pleistocene. These tools are for the most part not temporally diagnostic. However, immediately above Stratum 17, Valoch reports tools with affinity to the Szeletian, or early Upper Paleolithic.

Charcoal from a fireplace yielded uncalibrated radiocarbon dates of 32,990 ± 1500 BP (GRO - 848) and 32,420 ± 470 (GRO - 1724). These dates accord well with the Early Upper Paleolithic ascription of the artifacts found immediately above Stratum 17.

Pod hradem never yielded the substantial Paleolithic occupations that the excavators had hoped to find. Stratum 17 at Pod hradem is thus to all intents and purposes a paleontological locality.

Paleontological Summary

Musil (1965) studied the fauna from Pod hradem Cave. His main concern was the evolutionary trend evinced by the cave bears whose numerous remains were recovered from the locality. The reader is referred to Musil's contribution in the Pod hradem monograph for his detailed study of cave bear morphology and phylogeny.

Summary

The research results presented in the next chapters are aimed at better understanding the determinants of spatial distributions in pre-modern hominid archaeological sites, through systematic study of the traces of animal behavior. In order to achieve my research goals, I developed some quite selective criteria in the search for an appropriate faunal collection to examine. The Pod hradem Cave collection satisfies these criteria. First, there is no evidence for a hominid presence in the stratum chosen for study. Second, the cave was excavated over a wide enough area to allow statistical discovery and verification of spatial patterns. Moreover, recording of horizontal and vertical provenience was at a sufficiently fine resolution so as not to preclude the use of statistical spatial analyses. Third, data recovery was nonselective, resulting in the retention of all bone found by the excavators, undiagnostic and identifiable alike. Fourth, and perhaps most importantly of all, the cave's excavators granted me access to the collection, without reservation or restriction.

CHAPTER 4

Species Composition and Age Structure

By what processes did the bone specimens in Pod hradem Cave Stratum 17 come to be deposited? The site's excavators concluded that it was a cave bear hibernation den (Moravské Museum 1965), and similar inferences accompany numerous assemblages throughout Europe (Kurtén 1958, 1976). Although there are good grounds for crediting this conclusion, there are several reasons for further investigating the formation processes of the cave's bone accumulation. First, with the exception of a few pieces, the bones of the cave bears were not discovered in articulation. One might suggest that subsequent hibernating bears disturbed the bones of their predecessors in the course of preparing comfortable places to rest for up to six months. There is, however, ample evidence that other animals were involved in postmortem disturbance of the bears: a great many specimens have been modified by non-ursid carnivores. Lastly, bones of other carnivores occur in the assemblage, implying that the cave was at least a temporary residence for those animals. Therefore, it is at least conceivable that some or many of the cave bear specimens were transported to the cave by predators or scavengers, and that their horizontal distribution was affected by such carnivores. For these reasons it is necessary to apply taphonomic methods, with the goal of narrowing the range of processes that may have contributed to the bone accumulation. Only with such an investigation will it be possible to suggest how the Stratum 17 bones came to be distributed in space.

Taxonomic makeup of the assemblage, together with information about its age structure, proportional representation of skeletal elements, and modification of the bone, can be useful in determining the cause or causes of the bone accumulation. This and two subsequent chapters present the results of a taphonomic assessment of the Pod hradem Stratum 17 bone accumulation. After a description of field techniques and the condition of the fossils, this chapter examines the taxonomic makeup and age at death of the individuals whose bones comprise the assemblage.

Logistics and Field Technique

The Pod hradem Cave collection is housed in a depository of the Moravské Museum, in Budišov, near Brno, in the Czech Republic. I obtained permission to examine the specimens there from Dr. Luděk Seitl, then director of the Anthropos Institute. The Museum graciously accommodated me in a Baroque château, Zámek Budišov, in whose stable's attic the collection is kept. I was given a small table in a workspace in the collection room, in the eaves of the stable.

To obtain the results presented in this and following chapters I observed a number of physical characteristics on each specimen. All observations were made with the aid of a Donegan Optical Company OptiVISOR® Model DA-5, providing magnification of 2.5 X. For a lighting source, I used a harsh, 75 watt incandescent bulb. The OptiVISOR provided sufficient magnification to observe the bone surfaces for any meaningful alteration from the original bone morphology. All photographs were taken using an Ashahi PENTAX K1000 35mm SLR camera fitted with either a Tamron 28mm-70mm macro/zoom lens or Vivitar extension tube, with subjects lit by a 100 watt medium screw base light bulb for the primary, and reflected light for the fill source. Fine measurements were taken using a Ricoh digital calipers.

In the field, I recorded all observations on paper, and transferred them to electronic form once back in the U.S., using Borland Corporation's PARADOX® 4, a relational database management program. All statistical and spatial analyses presented in this study use data from this working database, and were undertaken using Statistical Graphics Corporation's STATGRAPHICS® *Plus* Version 5.

The catalog includes taxon, element, portion, side, age, sex, specimen completeness, articulation, conjunction, weight, length, width, cortical thickness, and categories of modification, as well as detailed data on the morphology of fragmentary specimens, and on location of modification. The complete catalog and data collection protocol, including computer codes and categories of observation, are available from the author upon request.

Condition of the Fossils

The fauna from Pod hradem Cave are preserved in an excellent state. Surfaces have undergone little postdepositional alteration, except near some carnivore tooth marks, where it is unclear whether cortical lamellae were removed at the time of gnawing or during subsequent movement, due to trampling or sediment compaction. The bones do not appear to be heavily fossilized, judging by their weight, although they have clearly absorbed minerals from the groundwater—most of the specimens are brown in color, with shades ranging from the color of teakwood through that of dark chocolate.

Curators applied a clear preservative coating at some time in the past, leaving all surfaces shiny, save those of the youngest individuals, which were the most porous bones. Most broken surfaces received the same clear coating as the exterior, leading me to believe that the majority of broken specimens came out of the excavation in the broken state. There is little evidence of curatorial damage or post-curation breakage. Clearly, however, some of the collection was damaged during excavation. "Fresh" chop marks and square punctures with yellow color exposed (?unmineralized bone?) attest to the use of picks and shovels during recovery. However, only 3% of the specimens display such damage.

Species Composition

Identifying the species making up a faunal assemblage allows one to begin assessing the processes that created it. Knowledge of the species present at Pod hradem will make it possible to say which species may have been living (and dying) in the cave, and which could only have been transported into the cave by scavengers or predators. Natural history and animal ethology are the principle theoretical guides in this endeavor. Knowledge of the habits of animal species, and of the variability in those habits, is helpful in beginning to comprehend the manner in which the bones of each identified species entered the cave. For example, bears, wolves and hyenas are known to den in caves. Others, like reindeer or bison, do not live in caves, nor would one expect to find parts of their carcasses deep inside a cave, unless they had been transported there, as carcass parts, by carnivores.

Method - Identification

Upon arriving at the Moravské Museum depository in Budišov, containing the Pod hradem collection, I discovered that specimens were stored in shoebox size and smaller cardboard cartons bearing arbitrarily assigned alphabetic designations corresponding to individual provenience units. These were stored in larger, numbered cartons. For identification and other tasks, such as determination of minimum numbers of elements and individuals, I needed to be able to remove specimens from their boxes and separate them from their box-mates, in order to put like with like. This meant that I first had to number each specimen with a unique identifier. I chose a trinomial that employed the larger storage carton's number, the smaller box's alphabetic designation, both of which might be shared with many other specimens, plus a unique sequential number assigned by me for each specimen. Thus provenience was encoded in the specimen number. I thought I had chanced upon an elegant time-saving procedure. However, it would appear that the elegance may have introduced too much complexity into the procedure. Upon my return to the United States I discovered that about 2% of the numbered specimens had been assigned duplicate numbers and thus had to be excluded from the study.

Fortunately, the errors were random. As such, these mistakes in no way jeopardize the statistical independence of the sample, nor do they threaten the plausibility of my conclusions. In a sample of over 2,000 specimens, the absence of a few dozen should not be a significant detriment.

Problems with the Identifications

The Pod hradem collection is housed in a depository of the Moravské Museum, at a distance of about 50 km from Brno, Czech Republic. There is no comparative collection at the depository, and the bones could not be moved to the museum, to enable comparison with its extensive fossil collections. I was thus forced to perform the bone identification using prior knowledge, photographs, and line drawings that I had brought with me from North America in anticipation of such difficulties.

If I had been able to employ a comparative collection, there is no doubt I would have made a great many more specific determinations.

Many of the specimens I could only minimally identify may ultimately have been added to the total for cave bears, if I had been able to eliminate other species. Without a comparative collection this was impracticable.

In spite of this, I feel that my inability to identify positively all of the specimens in the Stratum 17 collection docs not render my work invalid. Other workers, under optimal circumstances, have noted problems dealing with minimally identifiable skeletal elements and portions of elements (e.g. Driver 1982; Hesse 1982; Lyman 1979). Klein and Cruz-Uribe (1984:19-20) discuss the difficulty of assigning ribs to species, even when complete and in the hands of an experienced worker. They suggest, however, that it is defensible to assign a taxon to such hard-to-identify parts as sesamoids, ribs and fragmentary vertebrae using knowledge of the frequency of occurrence of a predominant taxon of appropriate size in the assemblage under examination. As it is, my ascriptions were conservative, in order to avoid any too tenuous conclusions. I decided instead to retain conservative taxonomic identifications, and note them where appropriate.

The task of identification was not always difficult in the absence of comparative material. There are so many cave bear specimens in the Pod hradem collection that I was able to find an example of every adult cave bear skeletal element, often from both sides of the body. This was enormously useful in identifying a great many of the specimens in the sample.

Unfortunately, there were no complete specimens of subadult elements and few complete neonate elements against which to compare the large number of immature cave bear specimens. Fortunately, however, subadult cave bears are sufficiently like their adult counterparts that in most cases they were readily identifiable. This was not the case for all subadult elements, as I relate below. I nevertheless feel confident of my identifications of subadult cave bear bones.

Identification of neonate cave bear specimens was a fairly easy task. They were uniformly dark brown and their texture, size, and morphology easily identified them as very young animals. Moreover, the vast majority were complete or nearly complete specimens. Although the morphology of the neonate skeleton was unfamiliar to me, it was still possible to identify elements positively using photographs, as well as the neonate's gross morphological similarity to more mature examples, e.g. position of nutrient foramina, and so on. I am relatively confident that my ascriptions in this category were also consistently accurate.

The most nettlesome identification task, and the one that, I fear, was carried out in the least satisfactory fashion, was that of the specimens I could not identify to a finer taxon than mammal. These consist mainly of long-bone shaft and rib fragments. Because of the inherent difficulty of identifying fragmentary long-bone shaft fragments and ribs, there are a number of minimally identified specimens, as will be seen in the discussion of results.

Some archaeologists might question the legitimacy of conclusions based on what appear to be somewhat tenuous identifications. A number of workers have made reference to the consequences of idiosyncratic data recovery, analysis and reporting (e.g. Payne 1975; Turner 1989; White 1992). At best, they aver, individual reports lack comparability with others because of a lack of specificity about what level of identification was employed—what is identifiable to one worker might be indeterminate to another. At worst, they argue, a great many archaeofaunas are of no comparative value: intersubjective variation renders each radically different from another.

White (1992:112-113) writes on the matter of the variability of expertise in identifying taxa and skeletal elements from fragmentary remains. Differences in ability and experience between and among faunal analysts and physical anthropologists, alike, present major impediments to comparative taphonomic studies (a point which has been made many times in the past by other workers). The problem of intersubjectivity has not, to my knowledge, been confronted, in public, with such vigor before White's contribution. He alludes to the verities of "hard" science, where measurement devices can repeatedly obtain precise, accurate estimates. Unfortunately for all of us, the human instrument, still all we have with which to perform faunal identifications, is notoriously variable. I concur with him that there is yet to be a solution, and I would suggest that there might never be.

If I were to take to heart such caveats and criticisms as those White justifiably voices, I might be tempted to seriously question the data I am presenting. I persevere only because I know that there have been others before me who have worked under similar constraints. The discipline of archaeology will have to come to terms with these issues sooner or later. I do not think, however, that the discipline will be served if scholars were to take part in self-imposed moratoria on faunal analysis, awaiting the time when they acquire the " 'feeling' or 'intuitive perception' [regarding bone identification] that no great paleontologist is without" (White 1992:113). The discipline must make do with what is has, while striving to improve the conditions under which work is carried out, and subsequently scrutinized by peers.

The following decision-making process governed my taxonomic ascription of non-ursid specimens. First, specimens identifiable to element were compared with outline drawings of the major Upper Pleistocene European fauna. Certain elements lent themselves more readily to specific ascriptions. Most, however, were only identifiable to Linnean family, and some, only to order. When I was unable to diagnose fragmentary specimens to high levels of specificity, I assigned them size classes within Mammalia. The majority of these pieces were shaft or rib fragments. Decisions about what size class in which to place a specimen were made by reference to the cave bear bones. Shaft cortical bone thicker than that occurring in adult *U. spelaeus*, I ascribed to very large mammal. Those similar in size to cave bear, I labelled large mammal. Those which more closely fit the model of juvenile *U. spelaeus* I placed in medium/large mammal, because these covered quite a range of sizes. Those that seemed smaller, I deemed small mammal.

Results

In all I examined 2,402 bone specimens in the sample from Stratum 17. There were originally 2,815 pieces in the collection, but I removed 365 fragments (mostly ribs) that looked like they had recent breaks. I did not want to include them because I could not use them in spatial analyses that dealt with specimen size. Moreover, because I could not be certain when the breaks had occurred, I was afraid any that had in fact been broken during or since excavation might artificially inflate the NISP for that element.

Taxonomic makeup of the Stratum 17 sub-sample is given in Table 4.1. Of the bone specimens, 61.7% are securely identified as cave bear (1,482 of 2,402). There are small numbers of other carnivores and artiodactyls, along with two bird specimens. A large canid, and a member of the genus *Crocuta* are also represented.

Interpretation

Table 4.1 reveals that about one-third of the specimens received only a general classification, such as large mammal. Although the 61.7% figure for the proportion of cave bear specimens in the sample is substantial, the actual number of cave bear specimens is probably much higher, given the conservative identification strategy outlined above. For example, of the 816 large mammal specimens, 316 are rib fragments, most likely cave bear, but which I did not feel confident

Table 4.1. Pod hradem Cave Stratum 17: number of identified specimens (NISP), by taxon.

TAXON	NISP	PERCENT OF ASSEMBLAGE
Artiodactyl indeterminate	18	0.7
Rangifer tarandus	5	0.2
Bovid indeterminate	1	trace
Rupicapra rupicapra	5	0.2
Capra ibex	1	trace
Carnivore indeterminate	5	.21
Ursus spelaeus	1482	61.7
Vulpes vulpes	1	trace
Canis lupus	7	0.3
Crocuta sp.	1	trace
Small mammal	3	0.1
Small/medium mammal	2	0.1
Medium mammal	1	trace
Medium/large mammal[1]	12	0.5
Large mammal	816[2]	33.5
Very large mammal[3]	2	0.1
Mammal, size indeterminate	38	1.6
Medium bird	1	trace
Medium/large bird	1	trace
TOTAL	**2402**	**n/a**

[1] Not *Ursus spelaeus*.
[2] *N.B.*: 318 of the Large mammal specimens are rib fragments, probably *Ursus spelaeus*, and at least 104 cranial fragments are probably those of juvenile cave bears.
[3] Not *Ursus spelaeus*.

assigning to that taxon in the absence of good comparative material. Most of these rib fragments are of a size that compares with others I was able to identify securely as cave bear, and most have the "wood grain" appearance and more-or-less square cross-section that characterize cave bear ribs.

In addition, a great many of the 104 cranial fragments in the category "large mammal" are parietal fragments from individuals of about the same age and species. I was unable to say with certainty to which taxon they belonged. However, given the large number of postcranial bones from cave bear subadults, I can say, with the appropriate caution, that it is likely these minimally identified cranial specimens, too, are ascribable to cave bear. I can thus tentatively suggest that the number of cave bear specimens could be as high as 2,000 (or greater than 80%), and perhaps more.

Hypotheses on the Origin of Species at Pod Hradem

In Table 4.1 we saw that, besides cave bears, the Pod hradem fauna contains a number of other taxa: birds and small mammals; very large mammals; carnivores, such as fox, wolf, hyena, and indeterminate; artiodactyls, such as reindeer, chamois, ibex, bovids, and indeterminate. It is possible to suggest some hypotheses as to how these various species might have entered the cave.

1) The bird, lagomorph, mammoth, pig, reindeer, bison, chamois and ibex are not likely to have entered the cave under their own power. It is therefore probable that these species were transported to the cave by scavengers or predators.

2) Hyenas, wolves and bears den in caves. Hyenas use caves when giving birth, and for some time afterwards, to protect their young from cannibalistic conspecifics and other predatory carnivores. They also use caves for refuge from competitors (Kruuk 1972). Wolves use caves for many of the same reasons. Hyenas commonly accumulate bones in their cave dens. The same is true of wolves. One would not normally expect to find the remains of hyenas or wolves in a cave other than one in which they denned or otherwise sought refuge. However, there is no way of knowing for certain if either species preyed on the other. Wolves and hyenas are no longer sympatric, as they seem to have been in Late Pleistocene Europe. It is possible that one scavenged the carcasses of the other, although the data to assess this possibility are presently unavailable.

3) Modern bears hibernate in caves, and it is thought, for a variety of reasons, that cave bears must have done so as well. It is thus likely that the cave bear bones are in the cave because they were dying there.

4) Both spotted hyenas and wolves are known to transport carcass parts of quite large animals—hyenas much moreso than wolves. Thus some, at least, of the cave bear bones might have been introduced when carnivores transported their bones from a remote death location.

Only the last two of these hypotheses are in competition. In order to narrow the range of processes potentially responsible for the bone accumulation at Pod hradem, it is possible to derive a set of expectations for what would occur in a cave that was used for different purposes by bears, wolves, hyenas, and foxes. Table 4.2 summarizes these expectations. If the behavior of cave bears was very different from that of modern cave bears, such that they did not hibernate, they might still have sought to give birth in caves or other protected locations, and to keep their young there until they were old enough to move about outside. In such a place, we should expect some infant mortality. Thus, any bone accumulation should consist of all the elements of very young cave bears. In a hibernation den, on the other hand, we should expect the same level of infant mortality, plus a number of the weaker, old adults that had succumbed to the rigors of infirmity and winter dearth. The natal dens of other carnivores should resemble a hypothetical cave bear natal den, consisting mostly of infants.

Foxes are not known to inflict serious damage to bone, through gnawing or disarticulation (Borrero 1990). However, Mondini (1995) documents their ability to accumulate the bones of small prey, such as rodents and birds, as well as those of scavenged camelids and ovicaprids, predominantly the small bones of the distal extremities. The bone collecting and modifying behaviors of wolves and hyenas are well known. Thus, fox, wolf, and hyena refuge dens should be expected to contain the bones of a variety of scavenged or prey species. Element representation would be the result of selective transport, and thus would not include all the elements of any taxon (except, perhaps, the remains of infrequent adult deaths of the species whose den it was). There would be a dearth of the very young of any species at such sites, since most would be consumed entirely, either in the open air, or in the shelter of the cave.

Finally, expectations for a natural trap are included, to provide an idea of what would be the case if animals were dying in the cave due to natural barriers to their escape. Such caves should contain the bones of a wide variety of species, of all age classes save, perhaps, neonates.

As I have suggested previously, there is good reason to believe that the bears hibernated in the cave. Since it is the task of this analysis to gather evidence about the processes forming the spatial distribution of bones in the cave, it is of paramount importance to understand the

Table 4.2. Characteristics of bone accumulations expected in cave bear natal dens, cave bear hibernation dens, wolf/hyena natal dens, wolf/hyena refuge dens, fox natal dens, fox refuge dens, natural traps.

	CAVE BEAR NATAL DEN	CAVE BEAR HIBERNATION DEN	WOLF/HYENA NATAL DEN	WOLF/HYENA REFUGE DEN	FOX NATAL DEN	FOX REFUGE DEN	NATURAL TRAP
TAXA	*Ursus spelaeus*	*Ursus spelaeus*	Wolf or hyena	Various	Fox	Various	All species
ELEMENTS PRESENT	All elements represented, within limits of recovery	All elements represented	All	Selective, based on prey size, age.	All	Selective, based on prey size, age	All elements represented
AGE CLASSES PRESENT	Predominantly neonate	High numbers of very young and very old	Predominantly neonate	Predominantly old adult	Predominantly neonate	Various	All ages
MORTALITY PATTERN	J-shaped age structure	U-shaped age structure	J-shaped age structure	U-shaped age structure	J-shaped age structure	Unknown	Living age structure

determinants of any spatial distributions. Thus, a thorough investigation of the characteristics of the assemblage must be undertaken, to determine whether the accumulation was the result of cave bear hibernation deaths, or of selective transport to a refuge or natal den of non-ursid carnivores, or some combination of those processes.

Non-Bear Taxa and Site Formation

The presence of carnivore prey species such as reindeer and ibex in Pod hradem Stratum 17, *and* the remains of wolf and hyena require that one investigate the possibility that Pod hradem was a den for other carnivores, and that they might have been the dominant agents of the bone accumulation. The presence of wolf and hyena may foretell considerable contribution on the part of these bone-collecting taxa. Hyenas, especially, are known to collect large numbers of bones in and near their dens (e.g. Kruuk 1972; Kerbis-Peterhans 1990; Lam 1992).

Non-bear carnivore taxa appear in low frequencies in the Pod hradem assemblage (refer to Table 4.1). It is a widely held notion that relatively high numbers of carnivore taxa in fossil assemblages identify hyenas as agents of accumulation (for a well-documented and wide-ranging analysis, see Cruz-Uribe 1991). Klein (1975), for example, observed that 22% of the total MNI were carnivorous species at Swartklip I in South Africa, a presumed paleontological hyena den. Scott and Klein (1981) found 31% of the assemblage at Deelpan, a modern South African hyena den, were carnivores.

However common it is to find high proportions of other carnivore remains at hyena dens, such numbers are not always the rule. For example, Hill (1989) found that only 9.4% of the bones at a recent spotted hyena den in Amboseli, Kenya were those of carnivores. Thus, although high numbers of carnivores in a bone accumulation may be used to tentatively identify hyena behavior, low numbers cannot be used as evidence of hyena's absence. The representation of non-bear carnivores at Pod hradem is therefore of dubious interpretive value. It may be true that other meat-seeking predators did not alternate occupations with hyenas (Stiner 1990a:684). If not, the few non-bear remains in the cave could be seen as the result of rare visits by carnivores during a hiatus in cave bear use of the cave.

Pod hradem's dental remains also provide evidence of carnivore contribution to the assemblage. Teeth are the most durable part of the skeleton under most conditions, and are the least desirable to predators from a nutritional standpoint, because they are almost entirely mineralized. They therefore tend to accumulate and to be preserved. If the teeth of non-bear taxa had ever entered Pod hradem, and had not been transported out in the guts of carnivores, they would most likely have survived in the assemblage.

At Pod hradem only six specimens of the dental sub-sample are non-ursid teeth (i.e. 0.025% of the total sample), while 214 (8.9%) are cave bear. In fact, only four of the six are from ungulates. In total 9.2% of the assemblage is comprised of teeth. Dental remains can be common in carnivore dens, at times occurring in numbers much greater than those present at Pod hradem (see Table 4.3). For example, at Swartklip I Klein (1975) reports high numbers of bovid mandibles and maxillae relative to postcranial elements in all but the largest animals. If even some of these tooth-bearing parts retained teeth, and this is a safe inference based on evidence from studies of the characteristics of hyena and wolf utilization (e.g. Binford 1981b; Haynes 1980), dental proportions would likely be much greater than the approximately 9% in Pod hradem. Lam (1992), in his study of variability in hyena behavior, finds that 40 of 445 specimens (8.8%) of a spotted hyena assemblage are isolated non-hyena teeth. Lam (1992) also reports that nearly 10% of the 445 specimens are mandibles and maxillae. If most of these retained teeth, the percentage of total teeth is also likely to have been substantially greater than the 9% at Pod hradem Cave. Behrensmeyer and Dechant Boaz (1980) report 20% (NISP) of the Amboseli hyena den assemblage is comprised of teeth, slightly less than the number left at death sites. And at a wolf den monitored by Binford (1981b), teeth also predominated.

Given the apparent propensity on the part of carnivores such as hyenas and wolves to collect tooth-bearing parts in their dens, the data from Pod hradem suggest that a low number of non-bear animal carcasses and carcass parts were brought into the cave by other carnivores. If such predators had contributed the remains of many prey individuals, and had themselves died in the cave in significant numbers, one would expect many more than 6 non-bear dental specimens.

A final class of evidence can be used to narrow the range of possibilities to explain the characteristics of the bone accumulation at Pod hradem, that of young animals. The only evidence of the young of non-bear species is a single developing enamel crown of a hyenid. Carnivore natal dens normally contain the remains of the young that succumbed during early life (e.g. Kruuk 1972; Stiner 1990a, 1994). These observations are for the most part anecdotal, and quantitative data are not easily available in the literature, so this characteristic will not be rigorously compared in this analysis. However, the near lack of

Table 4.3. Dental remains in carnivore assemblages, compared with non-bear and total dental remains from Pod hradem Cave Stratum 17.

	NISP	% NISP	REFERENCE
POD HRADEM STRATUM 17			
CAVE BEAR	214	8.9	infra
OTHER	6	0.25	infra
TOTAL	220	9.2	infra
AMBOSELI LANDSCAPE	5090	26	Behrensmeyer and Dechant Boaz 1980
AMBOSELI HYENA DEN	53	20	Behrensmeyer and Dechant Boaz 1980
KOOBI FORA HYENA DEN			
ISOLATED TEETH	40	8.8	Lam 1992
MAXILLAE AND MANDIBLES	32	>7	Lam 1992
SWARTKLIP			
MAXILLAE AND MANDIBLES	187	>19	Klein 1975
ITIKMALAIYAK WOLF DEN			
ISOLATED TEETH	35	31	Binford 1981b

evidence makes it hard to argue that the cave had functioned as a natal den for wolves or hyenas.

In summary, the preponderance of cave bears in the Pod hradem Stratum 17 bone accumulation suggests that they died in the cave. However, high numbers of cave bear specimens relative to those of other taxa do not provide sufficiently compelling evidence that the dominant mode of assemblage formation was autochthonous cave bear mortality. In fact, other lines of evidence hint at a more complex scenario. First, with the exception of rare portions of backbone the *U. spelaeus* specimens were not found articulated. This implies some active agents of disarticulation and dispersion. Moreover, remains of other carnivore species, as well as those of artiodactyls that could only have entered the cave as prey spoils, suggest that the accumulation is the result of many processes in combination. Finally, as I describe in Chapter 6, a great many of the cave bear bones show evidence of carnivore gnawing, some of it extensive. Thus, transport of cave bear bones into the cave by carnivores is a plausible alternative to death during hibernation as an explanation for the presence of *U. spelaeus* in this assemblage, and must therefore be investigated.

To be certain that the cave bears are not primarily the products of hyena or wolf predation and accumulation, other lines of evidence must be employed. A number of other parameters can provide insights into the agents of accumulation and disturbance at Pod hradem. The most important are: 1) the age structure of the non-bear carnivores; 2) the age structure of the bears themselves; 3) skeletal element frequencies; and 4) bone modification. Taken together, these can narrow the range of processes responsible for the bone accumulation. In the remainder of this chapter I will deal with age structure. In Chapter 5 I will discuss element frequency data, and in Chapter 6, bone modification.

Age Structure of the Pod Hradem Assemblage

Knowing the ages at death of the taxa in the Pod hradem assemblage can help to refine understanding of the likely uses to which the cave was put, and whether the species present were there because they were prey of other animals, or because they were living in the cave. Aggregates of these age determinations can tell us something about the way the animals died (see for example Klein and Cruz-Uribe 1984; Stiner 1990b). Frequency histograms of the number of individuals in each age class can be compared to the structure of a living population, to begin to assess the assemblage-forming process.

In most cases the frequency histogram of an assemblage will resemble either a "living structure" or a "U-shaped" mortality profile (see Figure 4.1). These correspond to the "catastrophic" and "attritional" profiles that one often encounters in the archaeological literature (e.g. Klein 1982). Terms such as "attritional," however, imply cause, which cannot be straightforwardly inferred from mortality profiles (Stiner 1990b). Thus, the more descriptive terms are here preferred. Stiner (1990b) has detailed the relationship between predator strategies and mortality patterns. Her insights inform the following discussion.

Both U-shaped and living structure mortality profiles can result from a number of different processes. If one were to track the prey ages of predators, different mortality profiles would be observed depending on the hunting strategy employed by the animal. Most predators fall into one of two classes, depending on their species-specific hunting strategy. Many large felids, for example, tend to take prey at random from a population, because they employ an "ambush" strategy. The death assemblage from this type of behavior would look like that of a living population, with differently aged individuals in numbers corresponding to their actual frequency of occurrence in the population. True catastrophes, such as floods, can also produce living structure death assemblages.

Wolves and hyenas belong to the group of "cursorial" hunters. These animals engage in long chases of their prey. Those targets less able to endure, usually the young and the very old, tend to succumb more often. Thus, the mortality profile of the victims of cursorial hunters will tend to resemble the U-shaped profile, having few prime adults. Cursorial hunters can, however, produce death assemblages that resemble those of ambush hunters, when circumstances (e.g. deep snow) give them the opportunity to bring down fitter prey. A U-shaped mortality profile can also result from true attritional deaths, when drought or starvation are the cause. In both these cases, the young and old are differentially affected, leading to a U-shaped profile.

Finally, in Table 4.1, the term J-shaped is used to describe the mortality pattern that would be expected in a natal den. There the rigors of early life would tend to skew the distribution strongly in the direction of the very young, given that neonate and young juveniles are susceptible to undernourishment, attacks by other carnivores, and so on (Stiner 1990a: chapter 13, 1994).

To summarize, bears hibernating, as well as wolf, hyena, and fox giving birth and rearing young, and fox, wolf, and hyena seeking refuge, are all known to encourage bone accumulations in caves and similar shelters. Knowing the age structure of the Pod hradem Stratum

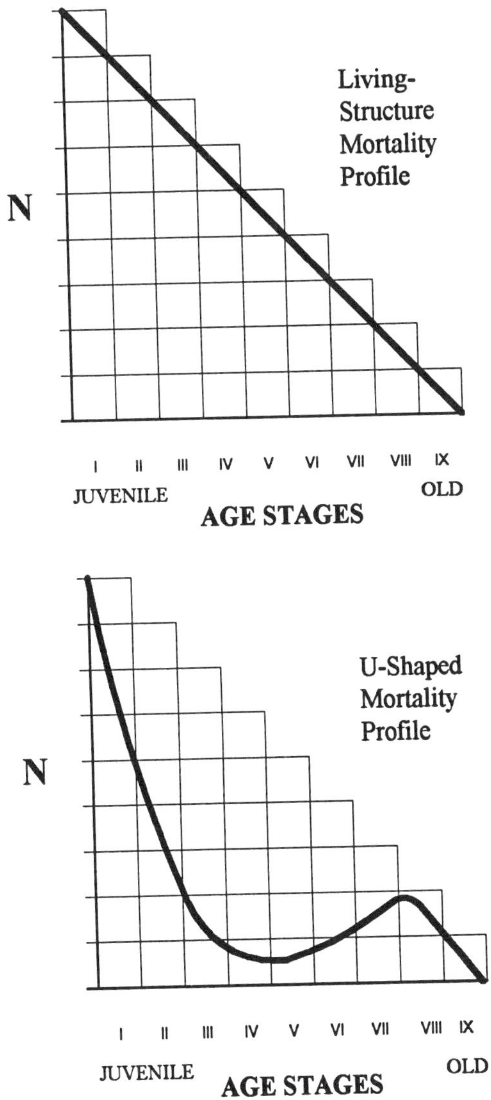

Figure 4.1. Idealized mortality profiles (after Stiner 1990b, figure 2, p. 310).

17 sample thus will help to narrow the range of processes likely to have been responsible for the bone accumulation at Pod hradem.

Method - Age Determination

Many elements of the mammalian skeleton can be employed to assess the developmental age of the individual. Most workers use dental development and attrition, epiphyseal fusion, and general morphological development. Sequences of dental eruption, attrition, epiphyseal fusion, and morphological change are fairly uniform in any given species. However, the age at which epiphyses fuse varies from element to element within a species. One must therefore know a great deal about the schedule of development of the species before epiphyseal fusion may be used to give age-at-death estimates. Without specific knowledge of the age at which fusions occur, all one can say of a bone is that it is immature or adult, depending upon whether the epiphyses are fused. I assigned an age to all *U. spelaeus* non-dental specimens using epiphyseal fusion, and morphological development. These criteria are sometimes subjective, as with color and texture of cortical bone, or size relative to the adult. Nevertheless, it was usually possible to assign a reasonably accurate developmental age to the non-dental bones.

For the mortality profiles given below, however, only dental development and attrition are used to estimate the age at death of an animal, because it is theoretically possible to obtain a calendrical age, rather than merely a developmental age at death. Enamel attrition can be used as a proxy for age, because animals that consume a great deal of vegetable matter usually wear down their occlusal surfaces at a relatively constant rate. This results in a regular sequence of morphological change throughout the life of a tooth.

Using known-age samples, one can construct a scale by which wear stages can stand as a proxy for age, either a calendar year, or an age class (e.g. young juvenile, old juvenile, subadult, adult, old adult). This technique has been used widely with hypsodont herbivores, and has also been applied with some success to low crowned animals such as pigs and bears (e.g. Grant 1982; Kurtén 1958). However, in the case of cave bear, no such known-age comparisons are possible.

I examined cave bear teeth articulated with mandibles and maxillae, as well as isolated teeth, using a descriptive system devised by Kurtén (1958) for cave bear dental development and attrition. He observed development and wear patterns on large samples of cave bear teeth, and noted that they tended to cluster into distinct wear-stage groups. Because they did not show a continuous series, he inferred that they must therefore represent mortality peaks occurring during late winter. Thus, all the bears that died during their first few months of life would possess teeth at roughly the same stage of development, given a relatively consistent birthing season occurring during hibernation. For older animals, clusters of similar wear stages probably also signalled annual mortality peaks. Working from a large sample Kurtén was able to describe up to seven age stages for some teeth.

Kurtén observed that cave bear cheek teeth began to develop early enough in life, and wear regularly enough, to provide a ruler by which to measure and monitor the age structure of the cave bear death assemblage. However, during my examination of Pod hradem Cave bear teeth in mandibles, I noted that a single individual could contain three cheek teeth which, according to Kurtén's system, would have estimated the age of the animal at three different stages. Thus, there are obvious problems with applying Kurtén's system to these materials in a straightforward way. To resolve this difficulty in the absence of modern, known-age comparative samples, I chose to look only at the tooth that occured most frequently in the collection, the upper second molar, in hopes of reducing the confusion that might occur if one were to look at the results of several different teeth using Kurtén's system. Using only one tooth, the groups I observed can provide a relative age estimate that still allows construction of a mortality profile of the Pod hradem Cave bears. Moreover, the upper second molar begins to develop in the first months of the cave bear's life, which means that even the earliest stage of life outside the womb can be monitored.

Kurtén's descriptions of the age categories for the upper second molar are reproduced in Table 4.4. In my analysis of the Pod hradem Cave Stratum 17 sample, I observed similar wear morphology to that observed by Kurtén, and was able, with reasonable ease, to assign his age classes to the sample.

Dental Remains

The cave bear dental age structure (Figure 4.2), based on 13 specimens, indicates high numbers of young (Stage I and II—i.e. around 0.4 and 1.4 years old), low numbers of prime aged animals (Stage III, IV, and V—i.e. 2.4, 3.4 and 4.4 years old), and relatively high numbers of animals older than stage six (Stage VI—i.e. 6.4 years and older). The result compares favorably with the so-called U-shaped mortality curve (see Figure 4.1). This mortality profile can be the result either of cursorial predation, or of death during hibernation due

Table 4.4. Kurtén's dental age criteria for the upper second molar (excerpted from Kurtén 1958:12-13).

WEAR STAGE	DESCRIPTIVE CRITERIA
STAGE I ≈ 0.4 YEARS	"The [converging proximal parts of the root walls are developed in many specimens, but some are even less advanced and consist practically only of the enamel cap] or slightly more advanced; the walls are formed somwhat less than halfway."
STAGE II ≈ 1.4 YEARS	"The roots are almost fully developed, but still open at the apices. The crown is unworn."
STAGE III ≈ 2.4 YEARS	"The roots are closed and the walls thickened. The crown is unworn or slightly polished on the summits of the internal cusps."
STAGE IV ≈ 3.4 YEARS	"The pulps are like pinpricks, and there is a very narrow attrition facet along the anterior part of the internal ridge."
STAGE V ≈ 4.4 YEARS	"The facet is somewhat larger."
STAGE VI ≈ 5.4 + YEARS	"The facet is more enlarged, but the tip of the protocone is still intact. The older specimens are lumped together for the present purpose."

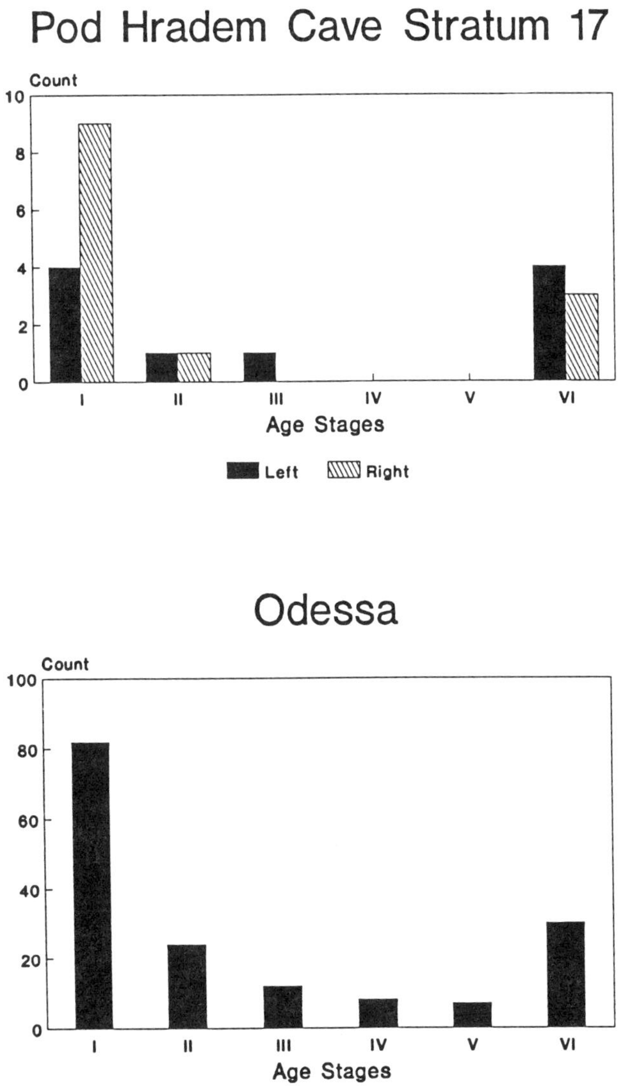

Figure 4.2. *Ursus spelaeus* mortality frequency histograms. Age determinations based on upper second molar, using Kurtén's (1958) descriptive criteria given in Table 4.4. Odessa data from Kurtén (1958).

to starvation or disease (which commonly affect the very young and very old).

Non-Dental Remains

Immature non-bear remains were almost entirely absent in the Pod hradem assemblage. By contrast, relatively high proportions of fetal and infant cave bear remains were present compared with their adult counterparts (Table 4.5). Approximately 25% of the identified *U. spelaeus* specimens (cranial and postcranial) from Stratum 17 are fetal or neonate. When subadult specimens are included with the fetal/neonate totals, fully 38% of the total NISP are from immature cave bears.

Interpreting Age at Death

From observations made on epiphyseal fusion and morphological development of *U. spelaeus* specimens, it is possible to conclude that high numbers of immature cave bears contributed their skeletal parts to the assemblage. Of the total number of specimens, up to 77% of some elements are from immature individuals (see Table 4.5). The average is closer to 38%, however, and the data require some further explanation. For some of the more durable elements, notably cranium, innominate, and the limb bones, juvenile MNE make up more than 50% of the sample. Elements such as the carpals, vertebrae and ribs, with lower percentages of NISP and MNE, are those most likely to have escaped the excavators' notice because they are too small in neonates, or, in the case of those elements having low volume density (especially the vertebrae), to have succumbed to postdepositional destruction due to trampling.

The data from dental attrition allow refinement of this dichotomous division between young and old. As Figure 4.2 illustrates, high numbers of juveniles and old bears characterize the assemblage, with almost no prime age adults in the sample. Such high numbers of juvenile and old adult remains would be expected in a cave bear hibernation den, where the rigors of winter would inevitably result from time to time in mortality among the weaker members of the population (Kurtén 1958). However, it might also have occurred due to cursorial predation, as noted above.

The data I have obtained for Pod hradem accord well with another locality inferred to have been the den of cave bears—a cave site near Odessa, in the Ukraine (Kurtén 1958). As with the Pod hradem

Table 4.5. Pod hradem Cave Stratum 17: cave bear age distribution.

	NISP TOTAL	NISP ADULT	NISP JUVENILE	% JUVENILE
CRANIUM	147	34	113	77
MANDIBLE	119	60	59	50
VERTEBRAE	189	162	27	14
SACRUM	7	7	0	0
RIB	191	167	24	13
SCAPULA	33	24	9	27
HUMERUS	60	27	33	55
RADIUS	50	16	34	68
ULNA	44	13	31	70
CARPALS	15	12	3	20
METACARPALS	28	27	1	4
METAPODIALS	17	4	13	76
INNOMINATE	21	5	16	76
OS PENIS	17	15	2	12
FEMUR	96	45	51	53
PATELLA	13	11	2	15
TIBIA	40	19	21	53
FIBULA	32	23	9	28
CALCANEUS	13	12	1	8
ASTRAGALUS	6	4	2	33
TARSALS	3	3	0	0
METATARSALS	29	29	0	0
1ST PHALANX	21	19	2	10
2ND PHALANX	7	4	3	43
3RD PHALANX	17	17	0	0
TOTALS	**1215**	**759**	**456**	**38**

N.B.: Disparities between Total NISP shown here and NISP in Appendix 1 and Table 5.1 reflect the omission of specimens whose age was indeterminate.

sample, I employed Kurtén's data on the upper second molar to produce the bar graph in Figure 4.2. His data were obtained using the same descriptive system of age estimation that I used. The mortality profile from Pod hradem and that derived from Kurtén's Odessa data (Figure 4.2) are almost identical. This strongly suggests that the same process was structuring both dental samples.

While disease and age may have taken their toll at any time, as with bears today, Pleistocene cave bears were probably more likely to succumb to old age, malnutrition, and disease at the metabolic low points in their annual cycle. From what we know of extant bears, these animals are at their most vulnerable during hibernation, especially females and their offspring (Kurtén 1976: 110; Rogers 1987). Those individuals unable to obtain enough calories during the summer and fall to build fat reserves are most likely to succumb to the rigors of hibernation (Rogers 1981). Offspring of undernourished individuals, and young in general, will be more likely to perish during hibernation, from disease, hunger, or both.

The evidence for seasonality inherent in the dental attrition categories is highly suggestive of hibernation death as the reason for the Pod hradem Cave bear accumulation. Indeed, as already noted, Kurtén (1958) interpreted such dental wear stages to represent annual mortality peaks occurring during hibernation. In the next chapter further light will be shed on the question of whether or not any of the cave bear individuals had been carried there as carcass parts by carnivores.

Summary

The Pod hradem cave bear bone accumulation has long been considered the result of bears dying during hibernation (Moravské Museum 1965). There is good reason to credit this conclusion. However, because the skeletal material was not articulated when excavated, and because of the existence of considerable modification to the bones by an unidentified actor or actors, it is necessary to investigate the possibility that it is a multi-actor assemblage, including transport of cave bear carcass parts by carnivores. Until it is determined what processes formed the site, one cannot be certain what caused any spatial patterning that might be in evidence.

The results of analyses of species composition and age at death suggest that Pod hradem may well have been a hibernation den. Dental attrition data suggest winter mortality peaks. Greater than twice as many cave bear juveniles are represented in the assemblage as adults.

However, the absence of prime age cave bear adults yields a U-shaped mortality profile, which could have resulted from winter (hibernation) mortality peaks in bear populations, or from the accumulation of carcasses hunted by cursorial predators. Low numbers of non-bear teeth, only one tooth of a juvenile non-bear carnivore, and overall low numbers of non-bear remains in Stratum 17 do not accord with the expectations of carnivore natal or refuge dens (see Table 4.2). Taken together, these data strongly suggest that during the accumulation of Pod hradem Cave Stratum 17 non-bear carnivores did not often use the cave as a natal den. In spite of this, it is clear that non-ursids used the cave sometimes as a refuge den. Thus, in order to more accurately assess the potential contribution of non-ursid carnivore transport through use of the cave as a refuge den, Chapter 5 documents the analysis of element frequencies and comparisons with modern carnivore open-air and den accumulations.

CHAPTER 5

Anatomical Composition

> Occasionally, to the delight of paleontologists, the entire skeleton of a long-dead animal is preserved, with every bone present and in place. Such an event occurs only in special circumstances: the body of the animal must have come to rest in a place where it could lie undisturbed for a great many years; it must have been rapidly covered with a suitable enclosing matrix, and it must have been preserved and strengthened by percolating solutions of appropriate chemical composition. More usually, the animal's body is subjected to the destructive influences that characterize any natural environment. The skeleton becomes disarticulated—broken down into its individual parts, each of which has then to contend with the attention of carnivores and with the forces of decay and disintegration.
> (Brain 1981:11)

Chapter 4 argued that death during hibernation was a likely explanation for the presence of numerous cave bear bones in Pod hradem Cave. However, there is still a question of whether or not some unknown number of the bear specimens were transported to the cave by predators or scavengers such as wolves and hyenas. This chapter analyzes the skeletal element frequencies of cave bear bones for what they can reveal about the processes responsible for the assemblage's anatomical composition. The first section presents an overview of the processes capable of altering the element frequencies of an assemblage, to introduce the conceptual framework within which this analysis of element frequencies was undertaken. This is followed by the results of my examination of the Pod hradem Cave assemblage. After that the Pod hradem assemblage is compared to some created by modern wolves and hyenas, the predators most likely to have impacted the study sample. This comparative analysis is useful for narrowing the range of possibilities, and tentatively suggests that wolves are more likely than hyenas to have been the primary modifiers of the cave bears' bones.

Determinants of Skeletal Element Representation

Only rarely does one find animal bone accumulations having the same numerical proportions of skeletal elements as those occurring in living animals: i.e. high numbers of teeth, vertebrae, ribs, bones of the manus and pes; low numbers of skull, limb, shoulder and pelvic girdle elements. There are many reasons for the infrequency of concord between natural and archaeological or paleontological assemblages. After a vertebrate dies, numerous processes can act to determine which, if any, parts of the skeleton will be represented in a fossil accumulation. Among the processes capable of perturbing the natural frequencies of elements are: differential destruction of bones of a certain size or durability due to weathering (e.g. Behrensmeyer 1975; Lyman and Fox 1989); consumption of the bone associated with desirable soft tissue by predators or scavengers (e.g. Brain 1981); dispersal and transport of carcass parts by those same predators or scavengers (e.g. Binford 1978a, 1981b; Kruuk 1972; Haynes 1980); transport of carcass parts by bone-collecting rodents, such as porcupines (e.g. Brain 1981); winnowing of lighter elements by flowing water (e.g. Behrensmeyer 1975; Hanson 1980; Voorhies 1969); displacement vertically in a soft substrate by trampling (e.g. Butzer 1981; Gifford-Gonzalez et al. 1985; Villa and Courtin 1983); comminution and eventual destruction of fragile elements by mechanical forces, such as sedimentary compaction or trampling (e.g. Shipman and Rose 1983; Behrensmeyer et al. 1986, 1989); recovery bias—failing to observe the smaller bones or choosing to discard parts deemed unidentifiable or unnecessary (e.g. Klein and Cruz-Uribe 1984; Payne 1975; Turner 1989; White 1992). The result of any or all of these processes in combination can be alteration from the expected frequency of elements comprising a bone accumulation.

Knowing the ways in which different processes can influence the anatomical composition of a bone accumulation makes it theoretically possible to infer the processes responsible for creating an assemblage from the frequency of its elements. I offer the following overview of some major determinants of skeletal element frequencies in fossil bone accumulations.

Cave Inhabitants

Animals that live in caves can also die in caves, contributing their bones to the cave's deposits. If a carcass is left undisturbed, anatomical representation should be identical with the animal's naturally occurring proportions. As already stated in the discussion of *U. spelaeus* in Chapter 2, they hibernated and probably died in caves. Extant hyenas and wolves also use caves, as refuges to which they transport prey carcass parts to get away from intraspecific competitors, and to give birth (Kruuk 1972; Stiner 1990a). Some caves are home to bats, owls, burrowing rodents, and the like. Thus, there is great potential for caves to accumulate the bones of vertebrates of all sizes.

Although the deaths of animals in the cave may explain how some of the bones might have entered Pod hradem, it cannot explain the final assemblage composition. Clearly the cave bears from Pod hradem are not represented by the proportions of elements occurring naturally in that species (see Table 4.5). At least one of the following may be involved.

Transport of Carcass Parts Away From Pod Hradem

Transport of carcass parts away from death sites by hominids and carnivores to other locales for consumption is a well-known phenomenon. This behavior explains, in large measure, many of the cave faunas that have come under scrutiny by archaeologists. Transport usually involves selection of carcass parts for reasons of utility, ease of transport, or end use. Thus, not all carcass parts are transported to a secondary location.

To enable identification of the processes that brought about the particular anatomical composition of an archaeofauna, archaeologists rely on comparative examples drawn from observations of wild and captive carnivores, and of humans living in traditional hunter-gatherer subsistence economies. These actualistic studies, to which I referred in Chapters 1 and 2, are not "merely a key to the past," as Simpson (1963:45) asserts, "they are all we have in the way of data."

Through actualistic study of living populations, it is becoming clear that animals and hominids contribute body parts to bone accumulations according to repetitive patterns of consumption behavior, whether this behavior is culturally mediated or not (see e.g.

Binford 1981b; Blumenschine 1986; Haynes 1980, 1983a, 1983b on animals, and Binford 1978a, 1984; Bunn et al. 1988; Gifford 1977, 1981; Gifford-Gonzalez 1991, 1993; Grayson 1988, 1989; Lyman 1985; O'Connell et al. 1988; Perkins and Daly 1968 on humans and hominids). Where unequivocally modern humans are concerned, there is considerable debate about the details, but it appears that situational decisions based on the utility and transportability of carcass parts, in the light of their intended uses, probably contribute most to variability in body part representation at archaeological sites.

Binford's (1978a, 1981b) ethnographic study of the Nunamiut of Alaska, an Inuit group whose subsistence focuses on seasonal caribou hunting, led him to conclude that there are variable, but nevertheless broadly predictable characteristics of bone assemblages left at procurement sites. These contrast with those transported to another location for consumption. Binford (1978a: Chapter 1) attributes this predictability to what he terms the utility of a given body part, in terms of its meat, marrow and bone grease content (which together form the Modified General Utility Index, or MGUI). Certain body parts are richer than others in these nutrients. Thus, when faced with decisions about what to expend energy transporting, the Nunamiut attached differential value to those parts containing the greatest amounts of nutritious tissue, while tending to leave behind those parts having the least nutritional value.

More recent research into human decisions determining bone assemblage composition at kill and consumption sites has revealed the complexity of human transport decisions in contrast to those made by animals. These decisions ultimately underlie the intersite variability observed in humanly produced faunal assemblages. Non-nutritionally based factors influencing modern human choices of what to leave at the dismemberment site of a large animal carcass include: energy constraints, such as distance from camp and the number of people available to carry meat and meat-bearing bone; social exigencies, including relationships governing carcass distribution at the death locale or at the destination; pragmatic considerations involving procurement and preparation techniques (e.g. Bunn et al. 1988; O'Connell et al. 1988; Gifford-Gonzalez 1993). These authors focus on different aspects—Bunn et al. (1988) and O'Connell et al. (1988) on the physical constraints placed on the process of transport, such as time of day, number of hunters available to carry carcass parts, and so on; Gifford-Gonzalez (1993) focuses on the important social relations of sharing and exigencies of food preparation. These studies have raised questions about the straightforward application of models based on nutritional utility to explain human carcass part transport. Instead, they demonstrate that strategic decisions regarding the disposition of carcass parts involve more than simple nutritional ranking of body parts.

These more-recent contributions may have the greatest impact on Binford's conclusions regarding economic utility and transport, since it means that humans will not make the same choices at all times and in all environments. Binford (1978a: Chapter 9) recognized the potential for these other factors to influence intersite variability. However, he continued to stress the interrelatedness of situational considerations and the underlying constant of nutritional utility.

With the accumulation of knowledge about the potential for different processes to create similar effects, archaeologists have become aware that they must first attempt to rule out other species' contributions to a given assemblage of bone in association with hominid stone tools, before ascribing the entire assemblage to hominids. This is the essential message contained in Binford's (1981b:26) introduction to *Bones: Ancient men and modern myths*, although he was referring to past processes in general when he said that archaeologists must first "attempt to isolate the different agents or forces that might be expected to contribute to or 'cause' a given pattern." Following that, archaeologists must develop criteria for recognizing the "traces, 'signature patterns' apt to be preserved in the archaeological record, of the agents likely to have contributed to the deposits." This, he said, must be done "by studying contemporary phenomena actively generated in a contemporary setting," because the object is to reduce the ambiguity caused by ignorance of the relationship between the causal process and phenomenon of interest in the present. Actualism, then, is the key to understanding past phenomena (cf. Gifford 1981).

In the service of reducing the ambiguity inherent in the traces of past hominid behavior, archaeologists have observed the behavior of carnivores and bone-collecting animal species in general, to gain comparative information on the character of assemblages produced by non-hominids. These studies have been primarily aimed at elucidating the behavioral evolution of hominids, specifically hominid meat procurement, whether by hunting or scavenging. However, they also provide baselines for understanding the formation processes of a wider range of paleontological and archaeological assemblages. Because the fossil evidence in Pod hradem points to the likelihood that wolves and hyenas contributed to the assemblage in one way or another, actualistic studies providing insights into the behavior of wolves and hyenas with respect to bone are most germane, and will be the only ones described here.

Haynes's (1982) rigorous observations of wild wolf behavior reveal a patterned sequence in the way they attack a carcass. This insight lies at the heart of the archaeologist's ability to characterize assemblages on the basis of their element frequencies. For, if carcasses are repetitively attacked in a certain way, then assemblages will tend to take on the shape of what is left after the carnivores have obtained all that they desire. Binford's (1981b) efforts to codify wolf and dog behavior with respect to bone makes a similar contribution.

Studies of spotted hyenas reveal that they can be prodigious collectors of bone. While their propensity to destroy bone in the act of consuming a carcass was already well known (Kruuk 1972; Sutcliffe 1970), their potential contribution to the archaeological domain was largely anecdotal until quite recently (e.g. Klein 1975). In a study that closely paralleled that of Haynes, Blumenschine's (1986) observations of the spotted hyena revealed a similar underlying logic to the content of hyena bone assemblages. Blumenschine maintains that this has the potential to allow archaeologists to assess the timing of hominid access to a given carcass. His "carcass consumption sequence," appears to be best explained by energetic and mechanical constraints placed on the carnivore, a point which echoes the findings of Haynes and Binford.

As Blumenschine observed, hyenas first remove all of the readily accessible soft tissue. After the easily removed soft tissue is depleted, the skeleton would then be attacked, mainly for the marrow it contained. The order in which the skeleton is reduced is best explained by the size of scavenged taxa relative to the consumer, and the consumer's difficulty of obtaining within-bone nutrients. For example, cancellous bone in long-bone epiphyses makes them much more susceptible to destruction by carnivores than the dense, compact, diaphyses.

Importantly, with regard to bone assemblage formation, all of the students of extant carnivore species mentioned in this section have noted that carnivore handling of prey varies as prey size increases. Each predatory species has a genetic predisposition to seek out meat and fat within the limits of their morphology. Even though hyenas, for example, are adapted to ingest and to break up bone, there is an upper limit to their ability to dismember, transport, enter the medullary cavity, and eat the bones of their prey. Thus, a hyena can access the brains of small and medium sized animals, but is apparently unable to break into the brain case of animals such as the buffalo or the giraffe. Likewise, wolves can access the soft tissue of the brain through the braincases of juvenile bison, but cannot do the same to robust adults, especially males (Haynes 1980). The same is true for limb bones as size increases.

Bones of juveniles are differentially more susceptible to modification or deletion by a consumer species than are adults, because development is incomplete among juveniles (see, for example, Gifford 1981; Richardson 1980). In general, young juveniles' bones are considerably more fragile than their adult counterparts because the bone is thinner and not fully mineralized. The susceptibility of young bone to destruction is exacerbated by unfused long bone ends, cranial bones, and so on, which renders them structurally weak as compared with the adults. Overall, their small size and structural weakness makes them easier to chew and break. This susceptibility to destruction may be to some degree balanced by the less attractive nature of the immature long bones, which do not store fat in their diaphyses in the way adult bones do. This preference for the diaphyses of adults has been observed in hyena and wolf consumption behavior (e.g. Blumenschine 1986; Gifford 1981; Haynes 1980; Richardson 1980). Variation in the degree of modification to bones belonging to animals of different ages is evidenced in Pod hradem's anatomical composition. Thus, differences in structure and nutritional value may help to explain differential survival of juvenile and adult cave bear elements.

Mechanical Forces

Bone is also susceptible to a number of forces and processes that cannot be attributed to purposeful animal behavior. These have received a great deal of research attention, because of the need to exclude their action in forming archaeological bone assemblages.

Voorhies (1969), Behrensmeyer (1975), Hanson (1980) and Badgley (1986) have all noted the tendency for bones of different density and shape to be moved differentially by flowing water. The densest bones are the last to be transported, resulting in a lag deposit of such heavy elements where water has passed over a surface accumulation of bone of different densities. Usually such assemblages are easily recognized. They are characterized by an overabundance of bones of a certain density or resistance to transport, such as teeth.

Trampling can also destroy and modify bone. It can fragment, and ultimately destroy all but the most durable elements and portions of elements. Heavily trampled assemblages retain few bones having low density. Finally, trampling also encourages downward movement of particles in soft substrates (Butzer 1981; Gifford-Gonzalez et al. 1985; Villa 1982; Villa and Courtin 1983). This has the potential to alter the horizontal integrity of an assemblage and could ultimately result in spurious element frequencies. It should be borne in mind that it is always a potential determinant of anatomical composition in an archaeological site.

Density-Mediated Destruction

The architecture of the vertebrate skeleton can be viewed as a response to the stresses placed on the bones by the animal's adaptation. Bone can be compact, and therefore dense, to provide maximum structural strength (e.g. long-bone shafts). It can also be spongy, to absorb shock and to minimize the overall mass of the skeleton (e.g. long-bone ends). Or it can be both at once (e.g. the cranial vault). Whether it is the jaws and teeth of carnivores, the pressure of hooves or paws, or rocks falling inside a cave, different parts of the skeleton (and the same parts of differently sized animals) will withstand those pressures differently, according to the relationship between total mass and total volume of a given segment of an element (Brain 1981; Lyman 1984). Because each element has a unique architecture, and each species has a unique morphology, different processes affect different bones differently across species.

Bone's differential susceptibility to mechanical forces is a function of which kind of bone is subjected to stress (i.e. compact, spongy, or flat). Research on this aspect of taphonomy continues to contribute insights into the formation of archaeological sites, even though the basic concept of density-dependent destruction has been known for some time (see e.g. Behrensmeyer 1975; Binford and Bertram 1977; Brain 1967, 1981; Gifford 1981; Grayson 1988, 1989; Lyman 1984, 1985).

For example, Brain (1967) catalogued the use of goat body parts, and their subsequent disposal, among a group of pastoralists in Namibia. He concluded that survival of goat parts is directly related to a measure of specific gravity (or bone density). The more resistant an element is to destruction, the higher will be the frequency with which it survives. Brain's data collection began in the 1960s. Since then, corroborating research was undertaken by Behrensmeyer (1975), and an exhaustive study of the subject has been contributed by Lyman (1984, 1985).

These studies demonstrate that there is a structured relationship between the differential destruction that occurs when animals consume other animals, and when any destructive process is applied to an assemblage of bones. To confound matters even further, it has been observed that often those skeletal parts deleted by scavengers or predators are the least dense, which has led to the suggestion that the less dense an element or portion is, the more likely it will be to succumb to destructive agents (e.g. Grayson 1988, 1989; Lyman 1985, 1992).

Considerations of element representation must always take place in the constant light of knowledge regarding density-mediated destruction. But whereas bone's density determines the result, it is still the interplay of behavioral or geophysical forces with the density of a given bone or segment that is most important.

Sampling Bias

Finally, I turn to a consideration of the archaeologist's potential to influence element frequencies, through bias during recovery or analysis. Selective recovery can either be conscious or unconscious. Even the most diligent excavator can miss the smaller specimens, unless sieves of appropriate mesh-size are consistently used. Those assemblages recovered without sieving, or using sieves of too great a mesh-size to catch small bones, will bias the sample against small elements and fragments. This usually means the omission of many microfauna, but it can also mean missing the smaller bones of fetal and neonate large mammals.

Often in the past an excavation was undertaken with the explicit intention to ignore animal bone that was not taxonomically diagnostic. This point has been made many times (see for example, Gifford 1981; Klein and Cruz-Uribe 1984; Payne 1975; Turner 1989; White 1992). Introducing such bias into analysis meant that limb bones, for example, might be underestimated, because of the tendency to discard shaft fragments which can be better estimators of element abundance.

One other aspect of recovery has been little considered (but see Villa and Delpech 1993), but may have an unforeseen effect. I speak of the partial excavation of depositional units whose contents are then used to form the basis of inferences about the formation processes of the unit as a whole. The complete excavation of depositional units in all but the most areally constrained of sites is a rarity. This may turn out to have been an unfortunate (if necessary) practice, since it can introduce unacknowledged bias to an analysis.

To date what has empowered archaeologists to dig a portion of a site and generalize based on the partial sample is the belief that statistical sampling theory is on their side (e.g. Binford 1964). The thinking goes something like this: if the process that created a bone accumulation also created a random spatial mixture of elements and sizes over the entire area of the accumulation, statistical theory tells us that a representative sample might be gained from excavation of only a portion of the area. However, the reality may be quite different. We know from experience with hominid and human encampments that things are often spatially patterned in nonrandom ways. And it is part of the purpose of this research to discover if the spatial distribution of bones in an animal cave is random or nonrandom. What does it do to our inferences based on partial samples if the sampling universe is not itself random? Clearly, it means that our samples are not representative of the sampling universe, unless they encompass that universe in its entirety—i.e. unless the entire depositional unit is excavated. Otherwise, we can never know what is in the unexcavated portions of a site based on what is found in the excavated portion of a site. This is a significant concern for those who study archaeofaunas from excavations that do not recover the entirety of a depositional unit. This probably describes the majority of published faunal assemblages, and certainly describes Pod hradem Cave, where the excavators purposely left one half of the area untouched.

An example from lithic studies will help to underscore and clarify this potential problem of bias deriving from the excavation of partial depositional units. Freeman (1992), excavating Cueva Morín in Spain, sampled the same depositional unit in two major episodes of excavation. Following analysis of the Mousterian lithics from the earlier excavation of Level 16 at Cueva Morín, Freeman declared that the assemblage was characterized as Typical Mousterian—i.e. it contained different kinds of Mousterian tools in proportions that resembled other assemblages defined as "Typical." Later, with the addition of lithics taken from the same level in a different area of the cave, the newly enlarged assemblage had to be characterized as Denticulate Mousterian—i.e. comprised of different proportions of distinct tool types, including a great many defined as "Denticulate." Great importance has been placed on such assemblage distinctions over the decades by Bordes (e.g. 1961a) and those who came after. Thus, Freeman's discovery of previously unacknowledged horizontal heterogeneity must strike fear into the heart of archaeologists whose inferences were based on samples drawn from vertical excavations, and not from excavations of entire depositional units. Moreover, faunal analysts who are used to making inferences of all kinds from similarly partial assemblages of bone, should take heed of the potential dangers inherent in these methods.

This example has served to demonstrate how tentative must be the conclusions based on assemblages drawn from partially excavated depositional units. Unless there is good reason to believe that the objects retrieved indeed derive from a spatially homogeneous distribution, there must always be space for revision. The same goes for those recovered using uneven size recovery techniques, and for any assemblage whose constituents were not retained in their entirety. This is a very tall order, and one that is not very likely to be satisfied in many cases. For this reason, conclusions based on such assemblages must always be treated as partial or preliminary.

Methods of Investigating Element Frequencies

Quantification

I first identified each specimen to as fine a taxonomic category as possible, using conservative criteria as outlined in Chapter 4. To gain an understanding of the processes contributing to the makeup of the assemblage I employed raw counts of bone specimens (NISP), an estimate of the minimum number of elements (MNE) and minimum number of individuals (MNI), as well as observed versus expected element frequencies (percent survival).

NISP includes all specimens identifiable to skeletal part, broadly defined (see Appendix 1 for the taxonomic and skeletal part inventory). MNE is an estimate of the minimum number of skeletal elements represented by the total number of specimens of each element. The latter statistic was produced by placing all specimens identifiable to cave bear and skeletal element together, enabling one to determine the minimum number of elements and individuals that could have contributed the observed specimens. In this determination, age, size, and morphology were considered, and MNE was derived from counts of the most commonly occurring portion of the element, once age had been controlled. Because I observed differential modification to the developmental age classes, most likely due to the huge difference in size between subadults and adults in the collection, I present MNE broken down into juvenile and adult categories.

The MNI statistic is an estimate of the minimum number of individual animals that must have contributed elements to the assemblage. I calculated it by obtaining the number of elements that can safely be said to be independent (e.g. not paired). This included age and size considerations, as well as epiphyseal fusion. Where all specimens were of one age group, the maximum number of specimens from one side of the individual was employed for paired bones, while for more commonly occurring elements, such as phalanges, the total number of occurrences was divided by the number of times that element occurs in the skeleton.

Percent survival, as explained in Brain (1981), reflects the relationship between observed and expected element frequencies, based on the highest estimated MNI for a given taxon in an assemblage. Percent survival provides a measure of relative skeletal element abundance, a way of quickly assessing the degree to which individual element abundance has been altered from the natural condition by taphonomic processes. For each element the expected number is calculated based on the frequency of occurrence in the animal's skeleton, and the observed frequency is expressed as a percentage of expected. So, for example, if two femurs survive from five individuals, the percent survival for femur would be 20%.

As mentioned above, numerous processes can bring about the anatomical composition of a given assemblage of bone, including destruction and transport of carcass parts by predators and scavengers, size-related sampling bias, incomplete excavation of depositional units, and postdepositional and diagenetic processes. Because it represents the cumulative effects of all of such processes, percent survival is not intended to reveal what Binford (1981b:26) called "signature patterns" of the agent creating an assemblage. Instead, it is a useful lens through which the analyst can see how the element frequencies of one assemblage compare with another, irrespective of assemblage size.

To test for statistical similarity between the Pod hradem sample and others I have employed only rank-order correlations. There is a sound theoretical basis for treating the data in this way. All of the parameters used for comparison (i.e. MNI, MNE, percent survival) are transforms of specimen counts and estimates, and are not, themselves, real things. They are thus best considered ordinal measurements (cf. Grayson 1984). Therefore, it is prudent to employ ordinal-scale statistics, so as not to violate any basic assumptions about the data that might be implied by using parametric statistics. This analysis employs both Spearman's and Kendall's rank-order correlations. Kendall's tau is a more accurate assessment of relationship between data sets containing a number of identical values (Blalock 1979:439), such as the juvenile component of the Pod hradem assemblage, which contains no specimens of many elements, and thus zero values for many categories.

In addition to their mathematical suitability to rank-order correlations, the causal processes underlying element survival in carnivore-kill assemblages are such that the data are naturally suited to rank-order comparisons. As Blumenschine and Haynes have observed, wolves and hyenas consume carcass parts in a regular order. This results in an intrinsically ranked representation of parts left at kill or consumption sites, such that the most readily accessible and desirable parts are deleted first, followed by successively less-desirable or accessible parts. Thus, it should be intuitively obvious that assemblages created by similar processes should look similar; those that are created by different processes, different.

Findings

Numbers of identified specimens, minimum numbers, and survival rates for juvenile and adult cave bear elements in Stratum 17 at Pod hradem are given in Table 5.1 (full skeletal-part inventories, by taxon, are given in Appendix 1). One of the first observations to be made is that every adult skeletal element is represented in the assemblage, while there are many significant absences among the juvenile specimens. Relative to the more-numerous cranial parts, more than 50% of the juvenile crania, mandibles, humeri, radii, ulnae and femora have survived. Juvenile innominates (21%) and tibiae (29%) did not fare as well. And juvenile vertebrae, ribs, podials, metapodials, phalanges, and other small bones have survived less than 10% of the time. Of the adult cave bear elements, mandibles, atlases, and axes have survived best, with survival rates of greater than 50%. Crania, vertebrae, ribs, long bones, and calcanei survive at between 25 and 50%. Only the hyoids, sacra, caudal vertebrae, sterna, innominates, carpals, tarsals, metapodials, astragali, and phalanges survived less than 25% of the time.

To better view the discrepancies between adult and juvenile survival rates, and to facilitate subsequent comparisons between numerical datasets, I have chosen to employ bar graphs based on percent survival. These allow one to view the "shape" of the assemblages at a glance. Figure 5.1[1] displays the percent-survival "profiles" for adult and juvenile cave bear from Pod hradem Cave Stratum 17. This figure clearly shows the difference between per-element survivals of juvenile skeletal parts and those of adults. More of the adult elements have survived, and most adult elements have survived at a higher rate than those of juveniles.

To assess the statistical significance of this observed difference between adult and juvenile percent survival, I first obtained the rank-order correlations between actual representation of elements (i.e. observed MNE) and the number that is estimated to have existed if no deletion of skeletal parts had occurred (i.e. expected MNE). Expected MNE was derived by multiplying the number of times a given element

Table 5.1. Pod hradem Cave Stratum 17: cave bear anatomical composition and percent survival of juvenile and adult specimens.

	NISP	MNE	JUVENILE		ADULT	
			MNE	% SURVIVAL	MNE	% SURVIVAL
Cranium	152	32	26	100	6	46
Mandible	122	54	41	73	13	100
Hyoid	18	2	0	0	2	15
Atlas	9	9	0	0	9	69
Axis	9	8	0	0	8	62
Cervical vertebra	23	20	2	1	18	28
Thoracic vertebra	95	81	3	1	78	43
Lumbar vertebra	37	35	1	1	34	44
Sacrum	7	2	0	0	2	15
Caudal vertebra	3	3	0	0	3	1
Sternum	6	2	0	0	2	8
Rib	202	156	5	1	151	42
Scapula	33	14	5	9	9	35
Humerus	62	44	35	63	9	35
Radius	50	39	34	61	5	19
Ulna	44	39	29	52	10	38
Carpals	15	15	3	1	12	7
Metacarpals	28	28	1	trace	27	21
Metapodials	17	17	13	n/a	4	n/a
Podials	1	1	0	n/a	1	n/a
Innominate	21	16	11	21	5	19
Os Penis	12	12	0	n/a	12	92
Femur	96	53	42	75	11	42
Patella	14	12	1	2	11	42
Tibia	40	27	16	29	11	42
Fibula	32	10	0	0	10	38
Calcaneus	13	13	1	0	12	46
Astragalus	6	6	2	2	4	15
Tarsals	3	3	0	0	3	5
Metatarsals	29	29	0	0	29	22
1st Phalanx	21	21	0	0	21	8
2nd Phalanx	7	7	0	0	7	3
3rd Phalanx	17	17	0	0	17	7

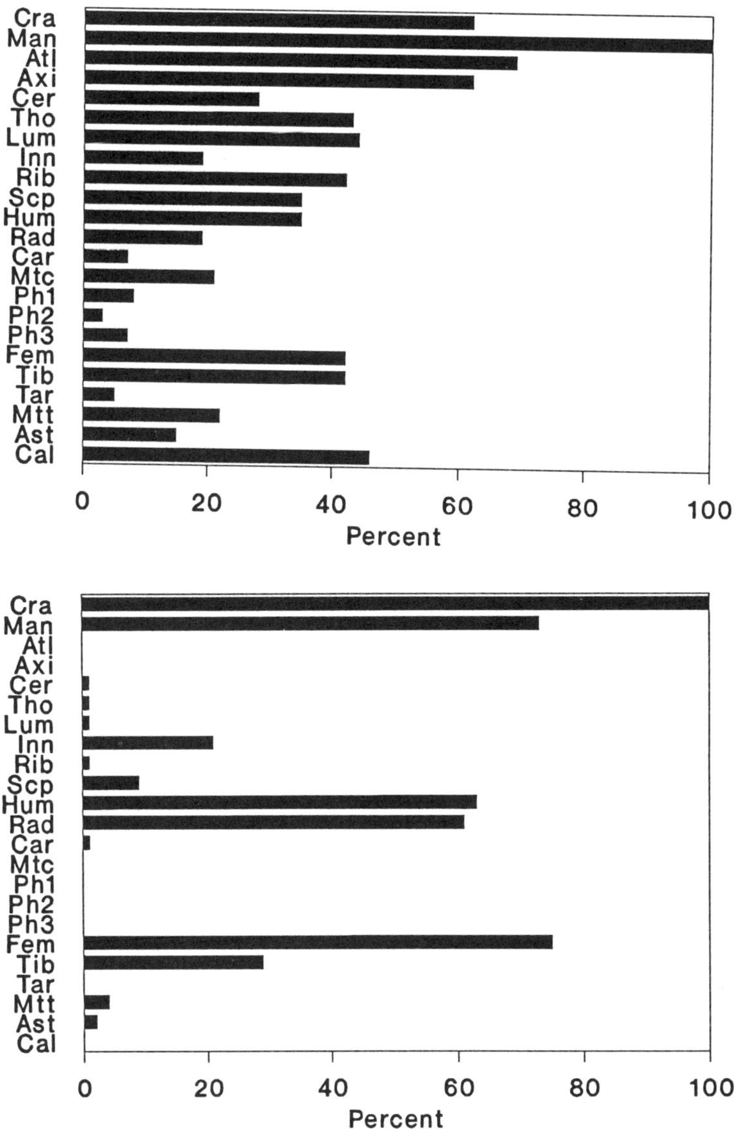

Figure 5.1. Percent survival of adult and juvenile cave bears from Pod hradem Cave Stratum 17: (top) adult; (bottom) juvenile.

occurs in the skeleton by the number of individuals estimated from the most commonly occurring element in the assemblage (i.e. MNI). Kendall's rank-order correlation between observed and expected MNE for adults is 0.3932 (p=0.0142, i.e. statistically significant). The same comparison for juvenile MNEs yielded a Kendall's rank-order statistic of -0.2906 (p=0.0790, i.e. not statistically significant). The difference between these two results reflects the structural difference between the two classes of specimen visible in Figure 5.1. Furthermore, when the survival percentages of juveniles and adults were compared I obtained a Kendall's rank-order correlation coefficient of 0.2043 (p=0.1879, i.e. weak and not statistically significant). Thus, whether comparing each age class with itself, or with each other, there is little concordance between these two datasets.

In summary, the relative abundances of juvenile and adult cave bear elements suggest that very different postmortem processes were at work causing the observed discrepancies in anatomical composition. These differential survival rates are most likely due to the disparity in size and durability between adult and juvenile bear bones, and the concomitant ability of destructive processes to alter element frequencies differentially across size ranges. The remaining sections explore further the observed differences between adult and juvenile element frequencies, and attempt to account for the anatomical composition of both subsets of the Pod hradem cave bear assemblage.

Accounting for Adult Cave Bear Element Frequencies

As mentioned previously, Pod hradem's cave bear bones displayed long-bone end deletion and high frequencies of gnaw marks. Such evidence of significant disturbance by carnivores compelled me to investigate their role in structuring the assemblage. A full discussion of modification to the cave bear bones, including an effort to determine the carnivore most likely responsible for the damage, will follow in Chapter 6. This section, however, assesses the impact of the numerous processes capable of altering element frequencies from their natural proportions, including those in assemblages produced by carnivores. It examines whether flowing water in the cave may have had an effect, as well as trampling, recovery size bias, and partial excavation of depositional units.

Assemblages reworked by water contain specimens of relatively homogeneous densities (e.g. Behrensmeyer 1975; Voorhies 1969), usually comprised of only certain elements. Which those elements are depends upon whether one is looking at bones left after being winnowed by flowing water, or those that have been redeposited after transport by a current of water. In the Pod hradem assemblage all elements are represented, from the densest, teeth, to what are probably the least dense, juvenile long bones. Thus, the range of sizes and shapes of bone specimens in Stratum 17 at Pod hradem Cave eliminates flowing water as a potential formation process.

If widespread and significant reduction of certain elements had resulted from trampling, the least dense parts of the skeleton would have been differentially destroyed. This does not appear to be the case at Pod hradem. While more fragmentary specimens do tend to occur in the excavation units nearest the cave entrance tunnel, the relationship between bone density and survival is not strong for the assemblage overall. I compared Lyman's (1984) bulk density figures from a deer skeleton with the survival rates of adult cave bear specimens. My use of deer as a surrogate for cave bear bone bulk density is motivated by the belief that the architecture of terrestrial quadruped mammal bones is more or less comparable from genus to genus—diaphyses are more dense than epiphyses, and so on. This use of analogous species is not without precedent. Grayson (1989) has used deer bulk densities to model the potential destruction of marmot bones. And, because I am using rank-order correlations, it is not the absolute values of density that are important, but the relative densities within each taxon that are important. A Spearman's rank-order correlation of volume density to adult *U. spelaeus* element percent survival yielded a coefficient of -0.0565 (p=0.8005). This demonstrates no relationship between what has survived and what should have survived if density alone were determining element frequencies. Thus, although trampling has had an effect on smaller pieces in the most heavily travelled part of the cave, mechanical processes are not, overall, responsible for the particulars of adult cave bear element frequencies.

Although the excavators were assiduous in keeping all bone specimens recovered, they may have missed the numerous, very small bones from fetal/neonate and subadult individuals. This cannot be ruled out as a potential influence on the element frequencies of those age classes. However, the overall size and durability of most adult elements precludes the possibility that recovery bias has had such an influence on their element frequencies. The modal specimen size was 40 mm. The smallest whole element in the bear skeleton would be the hyoid elements (about 30 mm) and the second phalanges (average 28 mm). Given that specimens only slightly larger than adult phalanges are the most commonly occurring in the assemblage, it seems unlikely that the adult sample has been affected by size-related recovery bias.

About one-half of the cave was left unexcavated at the level of Stratum 17. As mentioned above, archaeological statistical sampling strategies are born of the assumption that the sample will be drawn from a homogeneous random universe. However, samples drawn from heterogeneous distributions may not be used uncritically to form conclusions about the total distribution. Thus, there is no practical way to know for certain what element frequencies would be found in the unexcavated portion of Pod hradem Cave, without excavating it. The ambiguity that results merely forces me to be tentative in forming conclusions based on element frequencies alone. Thus, these conclusions must be viewed in concert with inferences drawn from the other lines of evidence presented in this study.

Pod Hradem Cave: A Scavenging Theater?

Several carnivores left their skeletal traces in Pod hradem Cave, among them wolf and hyena. Of the carnivores represented in Pod hradem, wolf and hyena are quite capable both of transporting and of chewing the bones of animal carcasses. Moreover, wolves have been known to prey on bears hibernating in dens (Horejsi et al. 1984; Paquet and Carbyn 1986; Rogers 1987:53; Rogers and Mech 1981). This implies that they could indeed have been active in creating the assemblage under study. Unfortunately, hyenas and bears are no longer sympatric species, thus it is impossible to know by analogy to present experience if hyenas preyed on bears in the Pleistocene. The present analysis has the potential to cast some light on this intriguing question.

Given knowledge of the fauna in central Europe during the Upper Pleistocene, and the skeletal evidence of wolf and hyena in the cave, it was necessary to determine whether Pod hradem could have been a site to which the cave bear bones had been transported by other large carnivores. I reasoned that it would be instructive to investigate the similarities (if any) between element frequencies of the Pod hradem cave bears and those of animals whose carcasses have been utilized by wolves and hyenas in caves and in the open air.

Transported Den Assemblages

All things being equal, one should not expect to find similarities between a transported assemblage and one comprised of what remains after carnivores have had their fill at a carcass. In order to demonstrate the differences, and to provide a database for comparisons with the Pod hradem sample, this section introduces some observations on transported wolf and hyena den assemblages. The following datasets have been used:

1) Kerbis-Peterhans (1990) investigated (among others) hyena transport behavior. Data on hyena transport for approximately bear-sized animals are summarized in Table 5.2. They derive from Kerbis-Peterhans's excavations of a striped hyena den near Arad, Israel (divided into camel and horse components), and a spotted hyena den excavated near Masai Mara, Kenya. Kerbis-Peterhans's data also include the spotted hyena den examined by Hill (e.g. 1989) in the Amboseli National Park, Kenya (Behrensmeyer and Dechant-Boaz 1980), a spotted hyena den assemblage recovered by Bunn (1983), and an aggregation of the *Equus* specimens recovered from all sites (Kerbis-Peterhans 1990:26, Table 2).
2) Binford (1981b:213) collected bone from the Bent Creek, Alaska, wolf den. These data are enumerated in Table 5.3.

Hyena den bone accumulations reflect a tendency for these animals to transport the more easily removed carcass segments, and to leave behind at the death site the spinal column and articulated ribs. Figures 5.2, 5.3, and 5.4 illustrate percent survival (based on MNI) of the six pertinent datasets drawn from Kerbis-Peterhans (1990). In all cases, vertebrae are scantily represented, as are ribs and the small bones of the lower limbs. Each of the bar graphs of these datasets has three peaks, where high frequencies of head parts, the bones of the forelimb, and those of the hindlimb are numerically predominant. In most cases, these peaks are of more or less equal magnitude.

Prior to using Kerbis-Peterhans's data for comparison to the Pod hradem anatomical composition, I subjected them to an internal check for similarity, employing the Spearman's rank-order correlation statistic. As Table 5.4 illustrates there is very high concordance, in all cases, among the hyena den assemblages. All are highly significant, with results ranging from 0.5663 to 0.8482, and are thus very robust results. This result accords well with Kerbis-Peterhans's findings, although he used a cluster analysis in his work. With so much similarity across space and species demonstrated by such high correlations, this augurs well for employing these data in a comparison with an assemblage of unknown origin from the past.

The homogeneously strong and highly significant correlations between and among hyena den assemblages is also good evidence of the efficacy of using rank-order correlations to compare element frequencies across assemblages. The only logical explanation for such concordance across hyenid species and across continents is the constancy of hyena behavior with respect to bone (a point made also by Kerbis-Peterhans with reference to these data). This level of

Table 5.2. Hyena den assemblage composition, percent survival based on MNI (after Kerbis-Peterhans 1990, table 2).

	AMBO I MNI=22	AMBO II MNI=11	ARAD I MNI=26	ARAD II MNI=25	SYOK MNI=11	MASAI MNI=24
CRANIUM	53	82	85	92	73	20
MANDIBLE	47	55	100	100	91	100
ATLAS	65	45	15	4	27	17
AXIS	24	9	8	4	0	13
CERVICAL	24	9	8	4	9	8
THORACIC	6	18	4	4	0	4
LUMBAR	6	18	4	0	0	0
PELVIS	65	27	54	32	18	25
RIBS	6	9	0	0	9	4
SCAPULA	65	36	4	40	27	46
HUMERUS, PR	18	18	0	36	27	33
HUMERUS, D	53	100	46	28	100	71
RADIOULNA, PR	71	82	35	72	55	96
RADIOULNA, D	53	46	23	48	27	46
CARPALS	12	18	4	4	9	38
METACARPAL, PR	53	64	38	64	36	58
METACARPAL, D	65	64	31	32	36	33
FEMUR, PR	18	18	8	16	18	38
FEMUR, D	35	27	31	36	9	29
TIBIA, PR	24	45	23	36	18	25
TIBIA, D	59	64	77	72	46	92
TARSALS	35	55	4	8	9	46
ASTRAGALUS	65	55	15	24	18	67
CALCANEUS	53	73	15	40	18	63
METATARSAL, PR	94	100	62	76	27	46
METATARSAL, D	100	91	54	28	9	25
1ST PHALANGES	24	18	4	0	9	0
2ND PHALANGES	6	9	0	0	9	4
3RD PHALANGES	18	9	15	0	9	4

AMBO I: Amboseli, Masai Mara, Syokimau *Crocuta* dens—*Equus* remains.
AMBO II: Amboseli *Crocuta* den—*Connochaetes* remains.
ARAD I: Arad, Israel *Hyaena* den—*Equus* remains.
ARAD II: Arad, Israel *Hyaena* den—*Camelus* remains.
SYOK: Syokimau, Kenya *Crocuta* den—Size III and IV Bovid remains.
MASAI: Masai Mara, Kenya *Crocuta* den—Size III and IV Bovid remains.

Table 5.3. Spearman's rank-order correlation coefficients comparing percent survival among modern hyena dens, based on MNI (data from Table 5.2, after Kerbis-Peterhans 1990).

SITE	ARAD I	AMBO II	ARAD II	SYOK	MASAI
AMBO I	0.6710 (p=0.0021)	0.7660 (p=0.0004)	0.7271 (p=0.0095)	0.5663 (p=0.0095)	0.6938 (p=0.0015)
ARAD I	-	0.7405 (p=0.0007)	0.8482 (p=0.0001)	0.6762 (p=0.0019)	0.6662 (p=0.0023)
AMBO II	-	-	0.8341 (p=0.0001)	0.6806 (p=0.0018)	0.7833 (p=0.0003)
ARAD II	-	-	-	0.7171 (p=0.0010)	0.8062 (p=0.0002)
SYOK	-	-	-	-	0.6404 (p=0.0033)

AMBO I: Amboseli, Masai Mara, Syokimau *Crocuta* dens—*Equus* remains.
AMBO II: Amboseli *Crocuta* den—*Connochaetes* remains.
ARAD I: Arad, Israel *Hyaena* den—*Equus* remains.
ARAD II: Arad, Israel *Hyaena* den—*Camelus* remains.
SYOK: Syokimau, Kenya *Crocuta* den—Size III and IV Bovid remains.
MASAI: Masai Mara, Kenya *Crocuta* den—Size III and IV Bovid remains.

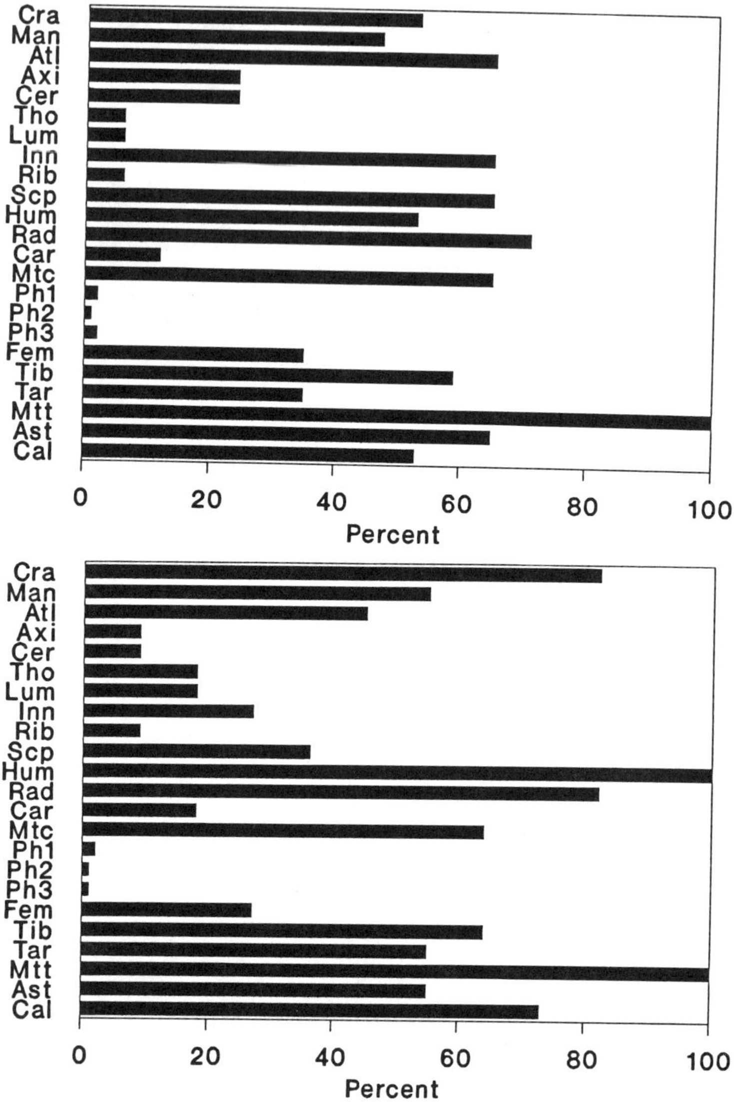

Figure 5.2. Percent survival of *Equus* and *Connochaetes* remains from *Crocuta* dens: top, *Equus* remains from Amboseli, Masai Mara, and Syokimau; bottom, *Connochaetes* remains from Amboseli (after Kerbis-Peterhans 1990:26, Table 2)

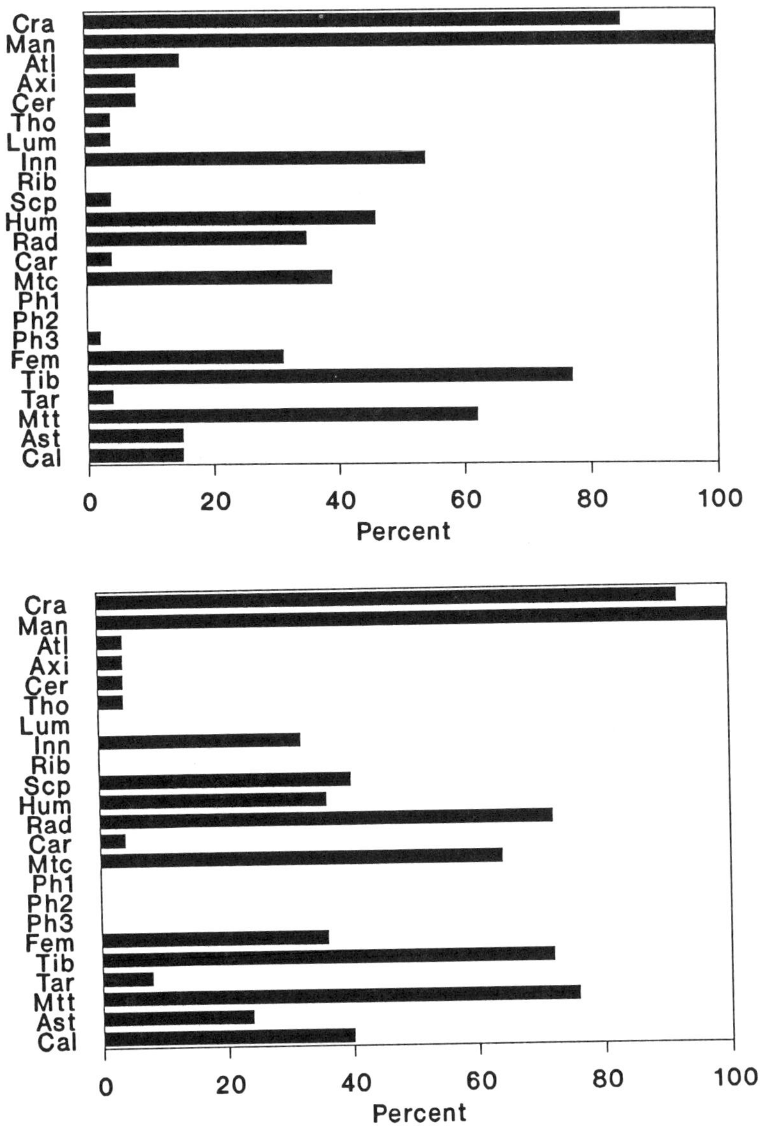

Figure 5.3. Percent survival of *Equus* and *Camelus* remains from *Hyaena* dens: top, *Equus* remains from Arad; bottom, *Camelus* remains from Arad (after Kerbis-Peterhans 1990:26, Table 2).

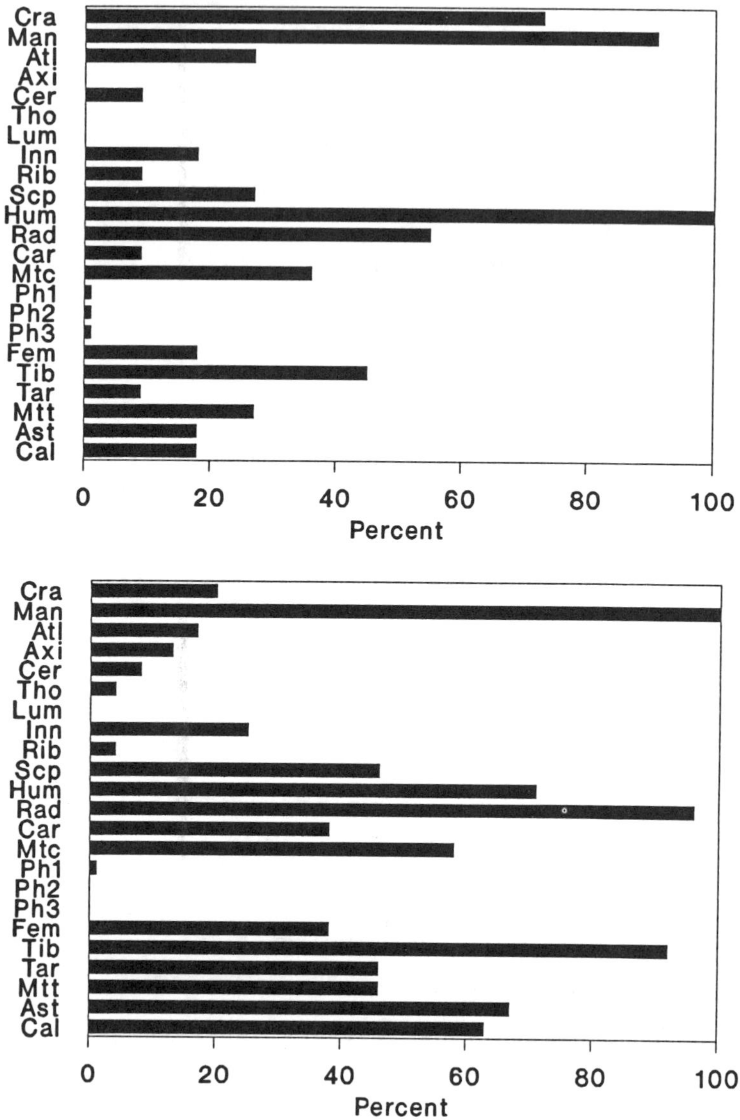

Figure 5.4. Percent survival of size III and IV bovid remains from *Crocuta* dens: top, Syokimau, Kenya; bottom, Masai Mara, Kenya (after Kerbis-Peterhans 1990:26, Table 2).

Table 5.4. Bent Creek, Alaska, wolf den assemblage composition, percent survival based on MNI of 14 (after Binford 1981b:213).

	PERCENT SURVIVAL
CRANIUM	86
MANDIBLE	89
ATLAS	66
AXIS	36
CERVICAL VERTEBRAE	6
THORACIC VERTEBRAE	2
LUMBAR VERTEBRAE	12
PELVIS (INCLUDES SACRUM)	68
RIBS	0
SCAPULA	57
HUMERUS, PROXIMAL	25
HUMERUS, DISTAL	100
RADIOULNA, PROXIMAL	75
RADIOULNA, DISTAL	46
CARPALS	32
METACARPAL, PROXIMAL	71
METACARPAL, DISTAL	68
FEMUR, PROXIMAL	14
FEMUR, DISTAL	21
TIBIA, PROXIMAL	14
TIBIA, DISTAL	61
TARSALS	36
ASTRAGALUS	32
CALCANEUS	36
METATARSAL, PROXIMAL	50
METATARSAL, DISTAL	29
1ST PHALANGES	23
2ND PHALANGES	16
3RD PHALANGES	7

consistency makes more warranted the application of such comparative data to analyses of Pleistocene bone accumulations, because it underscores the likelihood that hyena behavior is strongly patterned by a genetic makeup that is probably of great antiquity.

Figure 5.5 presents the percent survival profile for Binford's wolf-den data (1981b). The wolf-den profile is strikingly similar to those of Kerbis-Peterhans's hyena dens. Ribs and vertebrae arc rare or absent, as was the case in the hyena dens, in contrast to the relatively high survival rates of those parts in the Pod hradem Cave bears. It might be supposed that modern wolf data are potentially as reliable a model for Pleistocene wolf behavior as seems to be the case for observations of extant hyenas, although as wide a range of data as that available for the hyenas has not been collected for wolves.

A glance back at Figure 5.1 shows that the adult cave bear element frequencies from Pod hradem Cave Stratum 17 look substantially different from the internally consistent hyena-den assemblages. Cave bear head parts predominate, as do cranial elements in other carnivore den contexts, but in contrast to those other localities head parts at Pod hradem are followed closely in frequency by the robust atlas and axis. Moreover, the other vertebrae have good rates of survival, as do the ribs. Adult cave bear limb bones, on the other hand, appear at about the same rate as they would in a hyena den.

Table 5.5 displays the rank-order correlations between percent survival (based on MNE) of Pod hradem cave bear elements, and those from the cave sites monitored by Kerbis-Peterhans (MNE estimates based on MNI), as well as those of the wolf den observed by Binford. The Spearman's rank-order correlation coefficients are weak, and the levels of significance are, all of them, low. These correlations confirm that Pod hradem is in all likelihood not a concentration of transported bones.

These comparisons, plus the age-class frequencies discussed earlier, indicate that the Pod hradem assemblage was indeed most likely the result of its having been a cave bear hibernation locality, and not an accumulation of *U. spelaeus* bones transported there by other carnivores. In spite of the likelihood that the bears were dying in the cave, substantial modification to those bones indicates that carnivores were utilizing bear carcasses, and may indeed have been transporting bones *out* of the cave. This possibility is explored in more detail in the next chapter.

If wolves and hyenas were indeed consuming already dead animals in a manner typical of modern congenerics, a comparison with bone assemblages created by wolves and hyenas would be necessary to an understanding of the processes that formed the Pod hradem

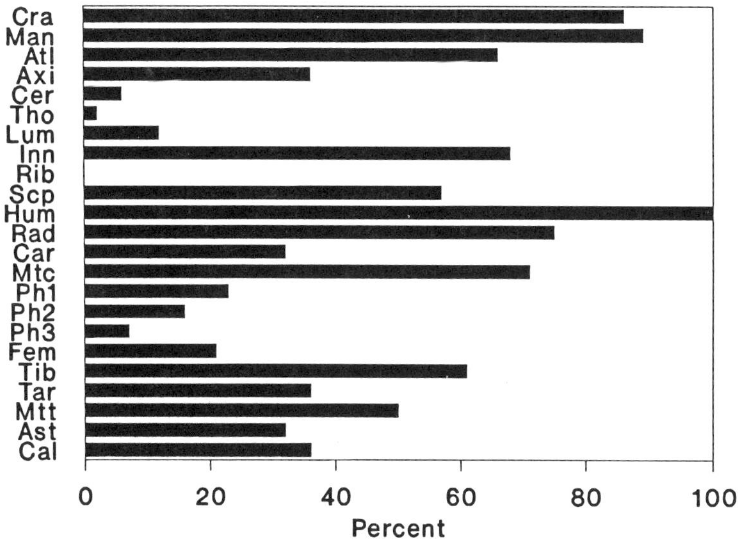

Figure 5.5. Percent survival of caribou remains from Bent Creek wolf den (after Binford 1981b:213, Table 5.01).

Table 5.5. Spearman's rank order correlation coefficients comparing percent survival, based on MNI, of modern hyena dens and Pod hradem Cave Stratum 17 adult cave bears (site data gleaned from Kerbis-Peterhans 1990: table 2; Binford 1981b:213; cave bear data from table 5.1, this dissertation).

SITE	POD HRADEM ADULT CAVE BEAR
AMBOSELI, MASAI MARA, SYOKIMAU CROCUTA DENS: *EQUUS* REMAINS	-0.0412 (p=0.8504)
ARAD, ISRAEL HYAENA DEN: *EQUUS* REMAINS	0.3158 (p=0.1479)
AMBOSELI CROCUTA DEN: *CONNOCHAETES* REMAINS	0.0735 (p=0.7362)
ARAD, ISRAEL HYAENA DEN: *CAMELUS* REMAINS	0.1766 (p=0.4185)
SYOKIMAU, KENYA CROCUTA DEN: SIZE III AND IV BOVIDAE REMAINS	0.0999 (p=0.6469)
MASAI MARA, KENYA CROCUTA DEN: SIZE III AND IV BOVIDAE REMAINS	-0.0466 (p=0.8311)
BENT CREEK, ALASKA, WOLF DEN: CARIBOU REMAINS	0.2456 (p=0.2493)

assemblage. Given that wolves and hyenas attack carcasses in a more-or-less-regular sequence, I would expect that they would attack carcasses in a similar way whether out on the tundra or in a cave (allowing for variation that might occur due to carcass availability, developmental age of carcass or carcass size). Thus, it should be worthwhile to compare modification to and deletion of skeletal elements by wolves or hyenas in the open air with what occurs in Pod hradem Cave. This section will look at wolf and hyena kill/consumption sites for comparisons with the characteristics of Pod hradem Cave's bear assemblage.

Open-Air Carcass Utilization

In the following analysis comparing bone accumulations resulting from open-air carcass utilization by hyenas and wolves with the Pod hradem assemblage, I employed the data below:

1) Hill (1980) enumerated a number of large (i.e. 200 to 300 kg) ungulate carcasses that had been left on the savannah in Amboseli National Park, Kenya, killed or scavenged by a variety of carnivores, including hyenas. Since hyenas are the only carnivores that readily chew bone, and are known commonly to remove carcass parts from the area of the carcass, most of the damage and survival noted by Hill is ultimately very likely to have been the work of hyenas. These data are summarized in Table 5.6.

2) Richardson (1980) collected observations of hyena-utilized size III ungulates (i.e. >300 kg). Richardson's data include buffalo, eland and giraffe. His data are presented as percent survival in the 1980 article, and by anatomical groups, not individual elements. As a result, I have had to extrapolate from the published data. This explains the numerous tied categories in Table 5.6 and in the percent-survival profile presented below.

3) Binford (1981b: table 5.01), while in Alaska, collected information on 24 wolf-killed caribou. I have transformed his Minimal Animal Unit counts to MNE for this analysis. Data are summarized in Table 5.6.

Figures 5.6 and 5.7 illustrate that all three open-air examples contain high numbers of head parts, as does the Pod hradem accumulation. In addition, these three present-day assemblages have high numbers of vertebrae and ribs, much more in keeping with the adult cave bear profile (Figure 5.1). The major difference among the four seems to be that the cave bear profile has lower percent-survival values overall, even though the shape of the profile is broadly similar to those of open-air carcass utilization. This subjective impression is

Table 5.6. Modern carnivore open-air carcass utilization: Binford's wolf-killed caribou (after Binford 1981b, Table 5.01); Richardson's hyena utilized size 3 bovids (after Richardson 1980:116, Figure 8); Hill's (1975) hyena utilized ungulates (after Kerbis-Peterhans 1990:26, Table 2).

	WOLF UTILIZED CARIBOU BINFORD (1981b) MNI=24			HYENA UTILIZED UNGULATES RICHARDSON (1980) MNI=7			HYENA UTILIZED UNGULATES HILL (1975) MNI=35		
	MNE OBSERVED	MNE EXPECTED	PERCENT SURVIVAL	MNE OBSERVED	MNE EXPECTED	PERCENT SURVIVAL	MNE OBSERVED	MNE EXPECTED	PERCENT SURVIVAL
CRANIUM	11	24	46			100	32	35	91
MANDIBLE	37	48	77			100	30	35	86
ATLAS	12	24	50			100	26	35	74
AXIS	17	24	71			100	34	35	97
CERVICAL	69	120	58			100	148	175	85
THORACIC	175	336	52			75	286	455	63
LUMBAR	78	120	65			70	145	175	83
PELVIS INCL SACRUM	44	48	92			N/A	35	35	100
INNOMINATE						65			
SACRUM						65			
RIBS	330	672	49			80	442	910	49
SCAPULA	24	48	50			80	58	70	83
HUMERUS						60			
PROXIMAL	5	48	10				28	70	40
DISTAL	19	48	40				38	70	54
RADIOULNA						60			
PROXIMAL	16	48	33				30	70	43
DISTAL	9	48	19				26	70	37
CARPALS	35	240	15			40	-	-	-

Table 5.6, continued.

	WOLF UTILIZED CARIBOU BINFORD (1981b) MNI=24			HYENA UTILIZED UNGULATES RICHARDSON (1980) MNI=7			HYENA UTILIZED UNGULATES HILL (1975) MNI=35		
	MNE OBSERVED	MNE EXPECTED	PERCENT SURVIVAL	MNE OBSERVED	MNE EXPECTED	PERCENT SURVIVAL	MNE OBSERVED	MNE EXPECTED	PERCENT SURVIVAL
METACARPAL						60			
PROXIMAL	11	48	23				26	70	37
DISTAL	11	48	23				20	70	29
FEMUR						60			
PROXIMAL	14	48	29				32	70	46
DISTAL	7	48	15				28	70	40
TIBIA						60			
PROXIMAL	10	48	21				34	70	49
DISTAL	22	48	46				44	70	63
TARSALS	40	192	21			40	128	280	46
ASTRAGALUS	11	48	23			60	28	70	40
CALCANEUS	11	48	23			60	30	70	43
METATARSAL						60			
PROXIMAL	17	48	35				34	70	49
DISTAL	9	48	19				28	70	40
1ST PHALANX	21	192	11			40	80	280	29
2ND PHALANX	15	192	8			40	80	280	29
3RD PHALANX	12	192	6			40	72	280	26

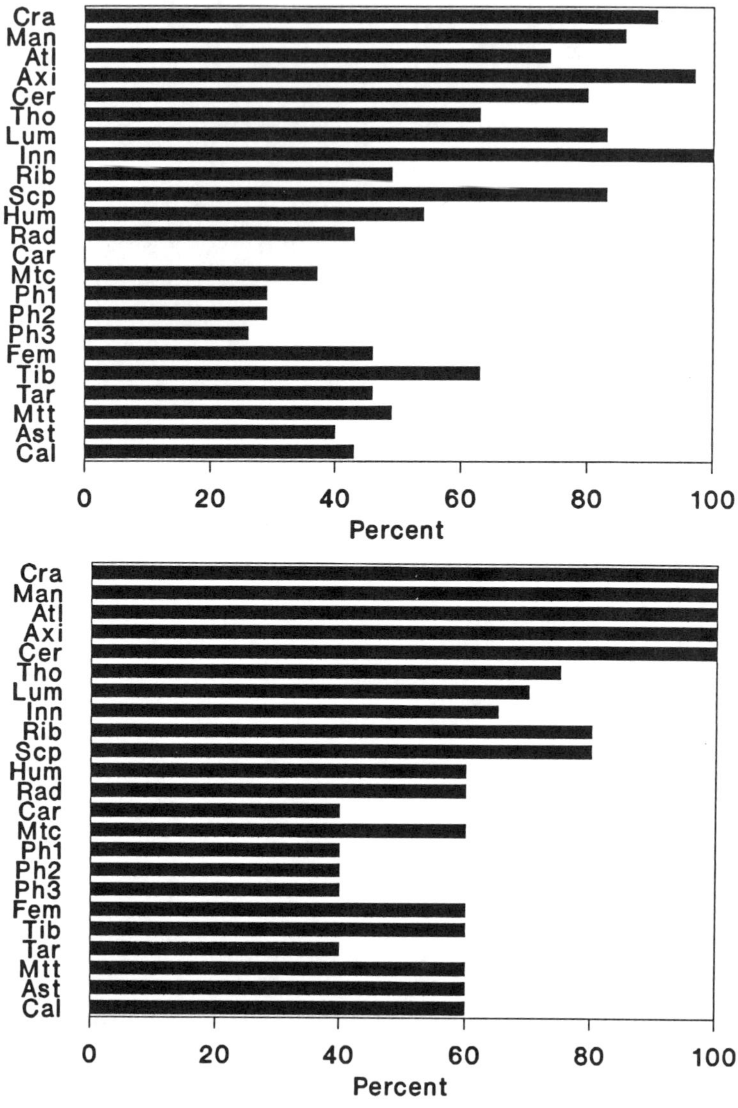

Figure 5.6. Percent survival of bovid and ungulate remains from open-air kill/consumption-site assemblages: top, size III bovid remains from Kenya and Uganda; bottom, size III hyena utilized ungulate remains from South Africa (bovid data from Hill 1975, after Kerbis-Peterhans 1990:26, Table 2; South African data from Richardson 1980).

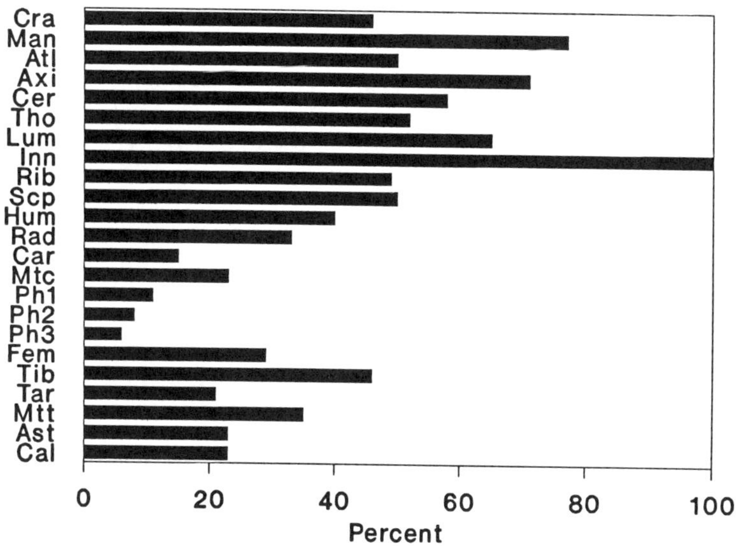

Figure 5.7. Percent survival of wolf-killed caribou remains from Alaska (after Binford 1981:210, Table 5.01).

bolstered by a relatively high statistical concordance between adult cave bear survival and those of Binford's wolf-killed caribou, Hill's landscape remains, and Richardson's hyena-utilized carcasses. A reasonably strong Kendall's rank-order correlation coefficient of 0.5473 (p=0.0003) was the result of a comparison of the adult cave bear percent survival and that of Binford's caribou. Moreover, the similarity between Hill's remains and cave bear results in a Kendall's rank-order correlation coefficient of 0.5282 (p=0.0005). These correlations are both strong and statistically significant. The Kendall's rank-order correlation between Richardson's ungulates and Pod hradem's adult cave bears is a less robust 0.3845 (p=0.0797, i.e. just short of statistical significance). This may be attributable to the method of data presentation and my extrapolation from it. It may also be that the small sample (n=7), and the larger animals included in the sample (i.e. giraffe), in some way contributed to the low correspondence between adult cave bear and Richardson's ungulates. In spite of this one, possibly spurious, result it would seem that there is indeed broad structural similarity between the anatomical composition of wolf-killed caribou assemblages, those of hyena-utilized carcasses, and of the adult cave bears at Pod hradem Cave. Yet, although broadly similar, there are some quite significant variances between the adult cave bear profile and those of open-air carcass utilization that warrant closer scrutiny. These are due in no small measure to some significant skeletal differences between ursids and artiodactyls.

For example, while high numbers of innominates characterize the hyena and wolf assemblages, cave bear innominata have survived in low numbers. This can probably be attributed to structural differences between cave bear and ungulate pelves. Whereas cow and deer innominates are quite flattened, and contain spongy bone only at the iliac crest and at the extremities of the pubis and ischium, the cave bear innominate is more like that of the human, with substantial regions of spongy bone within the compact bone of all three osseous components. This spongy bone would have contained desirable bone grease, and would have meant that carnivores attacked bear innominates more avidly than those of, for example, bovids or cervids. Being spongy, cave bear innominates would thus have succumbed to the carnivore's jaw forces. One additional factor may have led to the bear innominate's destruction—the region of the carcass containing the innominate is also a common access point for all carnivores consuming larger carcasses (Blumenschine 1986; Haynes 1982). Thus, not only is the innominate encountered early in the sequence of carcass consumption, but it is also quite desirable and easily consumed.

In addition to the obvious difference between bear and artiodactyl innominate survival, there appears to be a much-lower-than-expected proportion of cave bear lower limb elements, relative to that found in assemblages of hyena. Numbers of adult bear podials, metapodials, and phalanges are conspicuously low at Pod hradem, demonstrating a much closer concordance, in fact, with the survival rates seen at wolf kills of caribou (see Figure 5.7). Whether hyenas or wolves were responsible, there are at least two possibilities for such a dearth of paw bones.

First, the absence of the small bones of the feet might be the result of size-related recovery bias. There is, however, a very good reason to discount this possibility. The presence of a significant absolute number of these small bones in the Pod hradem collection attests to the fact that, although tiny, they were in fact frequently recovered. Although it cannot be assumed that uniform standards of recovery were maintained during excavation, it is nevertheless quite possible that the small elements of the extremities that were recovered represent an accurate estimate of the actual number originally in the deposits. Analysis reveals a direct correspondence between the size of foot bone and its rate of survival. A log-transform regression between the average length of element and its percent survival yields a correlation coefficient of 0.9700 (p=0.0003), an extremely high and significant result. Thus, the process responsible for the deletion of foot bones would appear to have been very sensitive to size differences. Although recovery size bias is a possibility, it does not fully explain the observations. The majority of bone specimens retained by the Moravské Museum excavators were on the order of 40 mm in length. Although the cave bear foot bones range in size from an average of 28 mm (2nd phalanx) to about 88 mm for the calcaneus, only the 2nd phalanx is small enough for it to have escaped the excavators' notice, if one assumes that the modal specimen length of 40 mm represents a threshold below which recovery was unreliable. Size-related recovery bias does not, thus, adequately explain the absence of larger paw elements, nor does it account for the almost perfect correlation between the size of the larger elements of the extremities and their percent survival. In this case, one must look to possibilities other than size-related recovery bias to explain the disproportional representation of foot elements.

Diane Gifford-Gonzalez (personal communication) suggested an alternative explanation for these patterns of element representation: bears are plantigrade, and their extremities contain fleshy pads, fat rich, that would be more attractive to consumers than ungulate extremities, as well as readily accessible to both large and small consumers. These contain the podials, metapodials and phalanges, all of which are relatively robust elements (and thus unlikely to have succumbed to trampling or other unintentional mechanical processes). Although large and robust, they would nevertheless have been easily ingested, either by hyenas or wolves. Binford (1981b:58), for example, found that wolves are capable of swallowing bones as large as 50 to 70 mm, which would include all adult cave bear foot bones, save the calcaneus. Hyenas, on the other hand, would have had no trouble with any of the bear foot bones. In Alaska wolves removed the small bones and in many cases swallowed them whole or in pieces, excreting them at their den (Binford 1981b:201). If this was happening in Pod hradem Cave, we should not expect to find small foot bones in the assemblage in high numbers. Rather, they would have been regurgitated or excreted elsewhere; perhaps in another part of the cave, perhaps not. The relatively few very small adult cave bear foot bones that remain in the Pod hradem assemblage may simply be those of individuals not heavily utilized by carnivores. Thus, in the absence of evident alternatives, it can be tentatively suggested that the low survival rates of foot bones have more to do with the behavior of modifying carnivores than with excavation technique. And it was the upper limit on the size of fragment that could be ingested by the modifying carnivore that most likely produced the very high correlation between element size and percent survival.

In sum, the Pod hradem adult skeletal element frequencies resemble most closely those of hyena and wolf open-air carcass utilization, with some indication that wolves were most responsible for the anatomical composition of the assemblage. But, were the bears being preyed upon? The answer is, "probably not." A hibernating bear would likely have presented a classic ambush hunting opportunity to the normally cursorial wolves and hyenas. Given that ambush predators prey on all ages, if hunted while hibernating, the mortality profile of Pod hradem's cave bears should resemble a living age structure. Instead, it is U-shaped, a profile usually associated with attritional deaths, such as would occur during hibernation. The conclusion to be reached, then, is that the bears were being scavenged.

Accounting for Juvenile Cave Bear Element Frequencies

The substantial difference between the shape of the percent survival profile for cave bear adult elements and that of juveniles (see Figure 5.1) results from low representations of all but the juvenile long bones and cranial bones found in numbers greater than about 20% of expected. This contrasts markedly with adult specimens, where only the innominate and the bones of the extremities survive in numbers less than 20% of expected.

There are several possible causes of these differences. Although carnivores have obviously modified juvenile bones, the most likely cause of the disparity in survival of adults and juveniles does not involve carnivore behavior at all. First, although it is clear that the excavators were assiduous in their collection, it is highly probable that many very small juvenile bear bones were missed in the excavation. This is especially true of fetal/neonate specimens—their feet bones are nearly microscopic in size. Second, the small size and delicacy of the majority of fetal or neonate cave bear elements renders them susceptible to all kinds of destructive processes, including comminution and subsequent omission during recovery. A neonate femur might be only 2.5 cm long, and composed of a thin layer (i.e. less than 1 mm) of cortical bone surrounding fragile cancellous tissue. An adult femur, on the other hand, is close to a half meter in length, with cortical bone thickness in places reaching 50 mm. Size must have played a role in survival—tiny phalanges, ribs, vertebrae, and parts of the axial skeleton of very small individuals would have been subject to differential destruction from trampling under the feet of 500 kg adults. Moreover, carnivores relatively incapable of modifying adult long bones could easily have swallowed fetal bones whole. Thus, Pod hradem was an inherently highly destructive environment for very young bear bones, with or without the addition of carnivore scavenging. With this in mind, further analysis is necessary to determine which process or combination of processes caused the disparity between percent survival of juvenile bones and that of adults.

To see if it was possible to discern any dominant process, I compared Lyman's (1984) bulk density values for the deer skeleton with percent survival of juvenile bear. If the survival of a given specimen depended solely on its density, as might be the case whether carnivores or trampling were responsible, there should be a strong positive correlation between density and survival. I obtained a Kendall's rank-order correlation coefficient of 0.2848 ($p=0.2028$), neither strong, nor significant. Even allowing for differences in the architecture of bear and deer bodies, this is a very low relationship between density and survival. This suggests that size-related recovery bias, rather than fragility of skeletal parts in the face of destructive processes, might better explain the under-representation of small juvenile parts.

Although carnivores no doubt had an effect on juvenile element frequencies, it would be superfluous under the circumstances to employ similar comparative tactics as those used in the previous sections with relation to adult cave bears. Given the high probability that juvenile element frequencies owe more to recovery bias than carnivores, comparisons with modern-day faunal accumulations could not add any finer resolution than that already achieved.

Summary

Anatomical composition provides another line of evidence in the search for the causal agent responsible for the characteristics of a bone assemblage. Adult element frequencies found at Pod hradem generally look much more like those of open-air wolf and hyena kill/consumption sites than those of their den sites. While on the surface paradoxical, the resemblance of a cave assemblage to that of open-air sites actually makes good sense. In Pod hradem, carnivores would have come upon the carcasses of recently dead bears and have utilized them according to the same habits typical of their handling of whole carcasses elsewhere on the landscape. By contrast, the contents of a carnivore den are normally bones which have been transported away from the kill/consumption site, creating a mirror image in terms of anatomical composition. Juvenile anatomical composition, on the other hand, appears to have been much more affected by size-related recovery bias.

Despite the close concordance between adult cave bear anatomical composition and that of open-air carcass consumption by wolves and hyenas, it will still be necessary to observe the patterns of bone modification to better understand whether either was primarily responsible for the character of Pod hradem's bone accumulation. That is the subject of the next chapter.

Notes

1. Abbreviations in Figures 5.1 through 5.7 are: Ast, astragalus; Atl, atlas; Axi, axis; Cal, calcaneus; Car, carpal; Cer, cervical vertebra; Cra, cranium; Fem, femur; Hum, humerus; Inn, innominate; Lum, lumbar vertebra; Man, mandible; Mtc, metacarpal; Mtt, metatarsal; Ph1, first phalanx; Ph2, second phalanx; Ph3, third phalanx; Rad, radius; Scp, scapula; Tar, tarsal bone; Tib, tibia; Tho, thoracic vertebra.

CHAPTER 6

Bone Modification

Chapter 4 presented compelling evidence that Pod hradem Stratum 17's bear bones had not been transported into the cave by carnivores. Chapter 5 documented that anatomical composition in Stratum 17 resembles that of open-air accumulations of wolf- and, to some extent, hyena-utilized carcasses. This chapter examines modification to the bones, i.e. evidence of the postmortem changes individual specimens have undergone due either to geological or to animal-behavioral processes. Ultimately it should be possible to determine if either wolves or hyenas were primarily responsible for the observed postmortem modifications, or whether in fact their "signatures" overlap to a degree that would make this distinction impossible.

It can be assumed that the cave bears, while alive, were capable of moving some of the bones at Pod hradem, and may thus have been responsible for the locations of an unknown number of the specimens recovered archaeologically. In other words, the bears very likely made a contribution to the spatial distribution of bone specimens. However, because it is also probable that both wolves and hyenas utilized the carcasses of the bears, it must be assumed that they were involved in creating some of the spatial patterning, even if it might never be possible to determine precisely which species was responsible for which patterns. If one or the other species of scavenger was the dominant modifying actor, it might be possible to ascribe certain patterns in the cave to that species. If, on the other hand, the evidence is equivocal, any patterning that appears might be attributed to a more generally carnivore level of behavior. Thus, it is important to extend the present examination of the bone accumulation, through a fine-grained analysis of the patterns of modification, to see if it is possible to assign priority to one or the other bone-modifying carnivore. In this way, the information at Pod hradem may indeed "open a window" on the propensity for either of those species, or carnivores in general, to create similar patterns in other cave sites.

Methodological Framework

Uniformitarian principles underlie this analysis of bone modification. To restate what was said in Chapter 2, the uniformitarian method involves recognizing structured similarities between the effects (i.e. the objects of study) of processes operating in the past and processes operating in the present that are known to produce the same effects. For example, knowing one or more causes for certain kinds of bone modification makes it theoretically possible to infer the process that modified bones in the past. One observes such parameters as element portion deletion or the surficial traces of modifying agents left on the bones. If a single animal, or species of animal, is responsible for the modification of a set of bones, if the conditions under which the carcass was utilized are known, and the effects of those conditions on the degree of utilization well understood, the archaeologist might have a very good chance of diagnosing the species responsible for the modification. However, one almost never knows the conditions under which an assemblage was formed in more than a broad sense.

Unfortunately for archaeologists, inferring processes responsible for ancient bone modification is not a straightforward endeavor. This is because different agents and actors in combination can produce similar marks on bone. For example, an agent such as a sharp-edged rock, if pressed against a bone by trampling hominids, could produce a mark similar to that of a hominid purposefully removing flesh from a carcass with a stone tool. Such potential ambiguity can leave archaeologists scratching their heads about how to distinguish among the candidates for assemblage "authorship." Difficulties are compounded when it is remembered that often more than one actor could have had an impact on a given bone accumulation—hominids and hyenas, for example. Ambiguity seems to be a symptom of the complex phenomenon under study, and something to be acknowledged and, if possible, overcome (Gifford-Gonzalez 1991).

As an example of the complicated inferential process involved where multi-agent assemblage authorship is indicated, Stiner's (1990a, 1990b, 1991a, 1991b, 1992, 1993, 1994) study of a number of cave faunas from west central Italy steers the most reasonable course through these murky waters. Stiner accepts the specter of ambiguity, and settles on what she considers to be evidence for a dominant actor in each depositional unit. This is accomplished through the application of multiple, independent lines of evidence, a strategy recognized as important to archaeological inference by Wylie (1989a) and Gifford-Gonzalez (1991). As Wylie (1989a:15) has said, "interpretive conclusions depend on various lines of argument ..., their strength derives not just from the diversity of their support but, more specifically, from the fact that the constituent strands draw on different ranges of background knowledge in the interpretation of different dimensions of the archaeological record; they are compelling taken together because it is highly implausible that they could all incorporate compensatory errors." Given that the present state of knowledge provides few categories of bone modification that might be used to unequivocally diagnose the actor responsible for a given trace on bone, Stiner's method seems most workable.

Having stated those caveats, this chapter on bone modification begins with an overview of the physical characteristics of bone modification—what happens to bone when it is affected by various processes. There follows a discussion of the actors and agents that can be responsible for bone modification. After that will be a short visit to the problem of teasing out the actors involved when dealing with multi-actor assemblages, and suggestions for dealing with variability that can arise from within-species differences in carcass utilization. Finally, I present the observations gathered during examination of the Pod hradem Stratum 17 assemblage, including my interpretations of the kind and degree of modification, and the tentative identification of wolves as the paramount assemblage authors.

Postmortem Modification

Postmortem bone modification can take many forms. Geological and biochemical processes can compromise bone's structural integrity, leading to volume loss and eventual destruction (e.g. Behrensmeyer 1975; Chave 1964; Miller 1975). Mechanical forces, such as saltation in a stream of water, or abrasion by wind-borne particles, trampling, or sediment compaction, can result in deformation, deletion of parts, and remodelling of a bone's surface (e.g. Behrensmeyer 1973, 1978;

Behrensmeyer et al. 1986, 1989; Miller 1975). Carnivorous animals are also capable of modifying bone to varying degrees (e.g. Binford 1981b; Haynes 1980, 1983b). A thorough assessment of the processes contributing to an assemblage of bone involves the search for evidence of geological and geochemical modification alongside the search for animal or hominid behavioral effects. I begin the assessment of bone modification by examining the effects of processes that do not involve animal modification.

Geological, Geochemical, and Mechanical Modification

Bone's susceptibility to disintegration in certain sedimentary environments is well known (e.g. Marshall 1989; White and Hannus 1983). An acidic sedimentary matrix can destroy bone, leaving only a difficult-to-monitor chemical trace (Chave 1964; Nichol and Wild 1984). Exposure to the sun and weather can hasten a bone's demise, through drying, leaching of the mineral constituents, and subsequent fragmentation (Behrensmeyer 1978; Miller 1975). Behrensmeyer (1978) observed patterns of weathering in aerially exposed bone in Sub-Saharan Africa, suggesting that there is a predictability to bone weathering. Lyman and Fox (1989), however, point out the variability inherent in the weathering process (e.g. due to heat and moisture [cf. Andrews and Cook 1985; Von Endt and Ortner 1984]), and caution against using Behrensmeyer's weathering stages as indicators of the length of time bone lay exposed prior to burial.

Bones that have been entrained in flowing water for a sufficient length of time can become rounded by abrasion against other particles, just as stones in creekbeds are rounded (Behrensmeyer 1982). Rounding that results from such sedimentary entrainment affects all projecting surfaces equally.

Compaction of sediments during diagenesis can result in plastic deformation, and postdepositional breakage of bone. I am unaware of any means of distinguishing between bones that have been broken by sediment compaction and those that have been broken in place by trampling on buried bone.

Animal Modification

There are two kinds of modification to bone that can be effected by animals: that which is intentional, and that which is the unintended consequence of other actions. The latter kind would include trampling. Trampling would no doubt have been a significant modifier of, especially, small, fragile bones in a cave such as Pod hradem, which was frequented by very large bears. The effects of trampling can range from sets of subparallel scratches on the surface of bones (see Behrensmeyer et al. 1986, 1989; Fiorillo 1989; Olsen and Shipman 1988) to fracturing and breakage (e.g. Agenbroad 1989; Bahn 1983; Haynes 1983a; Myers et al. 1980). Fractures and breaks are perhaps most difficult to ascribe to trampling alone, since sediment compaction can also produce breakage. For this study the bones from Pod hradem Cave were examined for scratches as described by Behrensmeyer et al. (1986, 1989) and Fiorillo (1989) for evidence of trampling damage.

Often during carcass consumption, carnivores will damage bones. Chapter 5 related how the redundant acts of carnivorous species on carcasses can result in element frequency signatures surviving at both kill/consumption locations and at places to which bones have been transported. The same logic of behavior applies in regard to bone modification. Static pressures produced by the action of jaw muscles can apply force at the tooth-bone interface, resulting in depression of, or removal of periosteal tissue and cortical bone lamellae, as well as crushing, fracturing, and separation of portions of the bone from the bone's main mass. The kind and degree of modification depends on the physical abilities of a carnivore species, the developmental age of the individual doing the damage, the size and developmental age of the carcass being utilized, and on the degree of utilization (e.g. Blumenschine 1986; Haynes 1980, 1982). Clearly, the importance of actualistic studies in producing knowledge of the possible effects of carnivores cannot be over-emphasized.

Due to the observations of archaeologists intent on understanding the processes of site formation, many animals are now known to modify bones. Predatory birds, such as owls and eagles, often swallow whole portions of their prey, later regurgitating hair and bones. Bones that have come in contact with avian gastric juices display characteristic modification, and are usually of a size and species that are habitually taken by certain kinds of birds (Brain 1981: Chapter 6; Dodson and Wexlar 1979; Andrews 1990). Bones modified by owls will come to rest beneath their roosts, and can form considerable accumulations, especially in caves and overhangs (Brain 1981; personal observation).

Other animals gnaw bone for various non-nutritional reasons. Rodents will gnaw degreased bone to keep their incisors occluding properly. Porcupine gnawing is eminently recognizable (Brain 1981; Shipman and Rose 1983). In addition, the porcupine is known to collect bones in its lairs, which can alternate with occupations by other species, further complicating the site-forming processes (Gifford 1981; Brain 1981: Chapter 5; Hughes 1961). Ungulates can modify bone in characteristic ways (Brothwell 1976; Gordon 1976; Haynes 1988a; Johnson and Haynes 1985; Olsen 1989; Sutcliffe 1973, 1977). Bears, too, can modify bone, leaving diagnostic gnaw marks (Haynes 1983b).

Field observation of carnivore behavior with respect to bone is an important contribution to the diagnosis of archaeofaunas. A number of workers have devoted their efforts to observing, describing and cataloguing the effects on bone of a range of mammalian predators, from foxes to hyenas (e.g. Binford 1981b; Blumenschine 1988; Borrero 1990; Brain 1981; Bunn 1983; Gifford 1977; Haynes 1980, 1982, 1983b; Marean and Spencer 1991; Marean et al. 1992; Miller 1969). In this way archaeologists are accumulating the tools with which to construct knowledge about the actors modifying bone assemblages.

I turn now to a consideration of difficulties inherent in dealing with multi-actor assemblages. Pod hradem Cave contains predominantly cave bear specimens, but it also includes those of hyena, wolf, and fox. All are capable of modifying bone. It must therefore be expected that the effects of a number of bone modifiers may be represented in the Pod hradem assemblage.

Multi-Actor Assemblages

It is probably the rare cave that has played host to a single bone-collecting or -modifying species (Farrand 1985; Ford 1976). Brain spent decades researching the processes forming bone accumulations at South African cave sites (see Brain 1981). His work emphasizes the taphonomic complexity possible in a single site. His actualistic research, observing the pellets of raptors, the carcass-utilization behavior of leopards, the collecting and modifying behavior of porcupines and domestic dogs (among others), has done much to both multiply and ameliorate the problems faced by the archaeologist when setting out to determine the actors responsible for a bone accumulation. In spite of the wealth of information he has gleaned on the characteristics of bones modified and collected by many species, Brain remains cautious about the ability of archaeologists to accurately assess the actors in multi-actor assemblages (see Brain and Turner 1984).

Nevertheless, there is cause for optimism, at least in the case of Pod hradem Cave, because of what is known about wolf and hyena bone modification. Binford (1981b) and Haynes (e.g. 1980, 1982), for example, investigated the behavior of canids in their carcass-utilization and bone-modifying abilities. Haynes (1980) found that wolves attack the carcasses of artiodactyls in a more-or-less predictable manner, foreshadowing the hyena consumption sequence observed by Blumenschine (1986), and the cumulative results of hyena transport behavior observed by Kerbis-Peterhans (1990). Habitual carnivore behavior is behind the characteristic element frequencies found in wolf-kill sites and reported on in Chapter 5, and it is what enabled Stiner (e.g. 1990a) to diagnose the bone accumulations of hyenas in Pleistocene deposits in west central Italy.

Thus, knowing the abilities of carnivore species to modify bone *can* enable an archaeologist to suggest the actor responsible for a set of traces on bone. This is especially true when the species differ in their ability to effect changes on the bone. Hyenas are prodigious modifiers of bone, and there is a quantum difference between the kind of modification a hyena and a wolf can accomplish, given a bone of a certain size. This, once again, means that it is theoretically possible to diagnose and distinguish between the modifications produced by hyenas and those left by wolves in the Pod hradem collection. There is, however, one problem that remains to be addressed: overlap between the modification seen on carcasses lightly utilized by hyenas and those heavily utilized by wolves.

Carcass Utilization Variability

Animal behavior is not invariable. Under different ecological circumstances, the same species will behave differently. For example, spotted hyenas have been observed to virtually ignore the bones of scavenged carcasses when animal flesh is plentiful, as in time of drought (Behrensmeyer and Dechant-Boaz 1980; Gifford 1977; Haynes 1980; Lam 1992). Predator population density also contributes to differences in hyena behavior (Gifford 1981), with higher densities leading to increased competition and more-complete utilization of carcasses by hyenas (Blumenschine 1986). Wolves, too, will only lightly utilize winter kills, because it is a time of relative ease of capture (Haynes 1980).

Observations of wolves suggest that they are more likely to utilize the bones of prey heavily at their "homesites," where adult and young spend more time "worrying" the bones (Binford 1981b:46; Haynes 1980). When prey are very large, the whole wolf homesite may move to a carcass, creating homesite utilization in the open air (Haynes 1982:268). Increased gnawing (leaving 75% of the bones toothmarked, and break margins rounded by licking and contact with sediments) is one of the results of the entire home group utilizing a carcass. Variability in the intensity of carcass utilization means that carcasses little utilized by hyenas (the more-capable bone modifier) may look as if they had been heavily utilized by wolves (the less-capable modifier) (Haynes 1980, 1983b). This could prove to be a nettlesome problem at Pod hradem, where bones of both species are found along with those of the cave bears.

Finally, the ability of a species to modify bones depends on the force that they can produce between their jaws, the strength and morphology of their teeth, and the strength and morphology of the bone being attacked. The bigger or more dense a bone is, the less susceptible to carnivore modification it will be (e.g. Brain 1981:19; Lyman 1984, 1985). Wolves, for example, can devour the crania of 150 kg juvenile bison, whereas they have less success with robust adult crania (e.g. Haynes 1988b:230). Conversely, the smaller, or less developed a bone is, the more likely it is that the bone will be demolished by a carnivore (e.g. Blumenschine 1986). Richardson (1980) notes that subadult individuals of prey species are especially susceptible to destruction because their epiphyses are unfused, rendering the long-bone ends structurally unstable.

Blumenschine (1986:647) adds that a given bone-modifying species will attack the carcasses of differently sized prey differently, according to the availability of within-bone nutrients. Hyenas will also attack the carcasses of juveniles differently than adults of the same species, due to the desirability of the within-bone component of juvenile tissue. Indeed, the bones of subadult individuals are often less attractive, because their marrow cavities are the site of red blood-cell production, and not the repositories of fat that they are in adults (Blumenschine 1986).

In summary, there are a great many ways that bone can be modified, and a great deal of variability inherent in the way carnivores modify it. Observation of variable behavior on the part of a number of species leads to the conclusion that the task of diagnosing actors in the formation of an animal bone accumulation is complicated along many dimensions. Carnivores having different bone-modifying abilities had access to the bears in Pod hradem Cave, creating the potential for a very ambiguous set of traces on the bones recovered from it. And, because almost all stages in the development of cave bear are represented at Pod hradem, one can expect to observe the effects of even a single species of bone modifier treating the bones of differently aged cave bears differently. If Pod hradem contains any coherent information about the ability of any one species to create spatial patterning, it will be necessary to determine whether or not one species' behavior is primarily responsible for the damage and disturbance observed in the assemblage.

Results

Geological, Geochemical, and Mechanical Modification

None of the specimens in the Stratum 17 sample showed signs of weathering as defined by Behrensmeyer (1978), and all were exceptionally well preserved. Although many specimens bore fine scratches on the compact bone, I was unable to find the sets of subparallel scratches attributable to trampling, as described by Behrensmeyer et al. (1986, 1989). However, I sometimes noticed that fine scratches occurred on the most salient parts of long bones, perpendicular to the long axis. These I attributed to trampling. Perhaps because of the predominantly fine-grained sedimentary matrix, cortical scratching from trampling was minimal.

Due to the normally high levels of calcium carbonate in a limestone cave such as Pod hradem, bone usually preserves well, as I explained in Chapter 2. In addition, weathering such as that described by Behrensmeyer (1978) normally occurs in the open, where sun and moisture can work to destroy a bone's integrity. Thus it is not surprising that the bones from Pod hradem should show no evidence of modification due to geochemical processes.

The Pod hradem Stratum 17 collection shows no certain evidence that it has undergone significant geochemical, geological, or mechanical modification. However, almost 6% of the specimens show overall rounding. Trampling is the most likely the ultimate cause of this modification, with repeated pressure against encasing fine sediments the probable proximate cause of the modification.

Carnivore Modification

This discussion of the effects of mammalian carnivores on the Pod hradem Stratum 17 assemblage is only possible because of the careful field observations of a growing number of archaeologists concerned with expanding the knowledge of animal behavior with respect to bone. One in my position can only hope to tentatively identify the marks of carnivore utilization by reference to published descriptions of individual effects and patterns of modification, and by examining, where possible, the collections of those who have either experimented with carnivores and their effects on bone, or who have collected carcass parts from known-actor episodes in the field. Dr. Gary Haynes, of the Department of Anthropology, University of Nevada at Reno, graciously afforded me the opportunity to examine his collections of bear-, wolf-, and hyena-gnawed bone before and after my examination of the Pod hradem assemblage. My assessment of the actors that modified the Pod hradem assemblage was greatly facilitated by visits and conversations in Dr. Haynes's laboratory.

Animal Gnawing

I inspected all bone surfaces under 2.5 X magnification. Any modification to the bone's surface that involved removal or displacement of subchondral bone or cortical lamellae, leaving a mark that is u-shaped in cross-section, and either circumscribed, linear or sublinear in plan were recorded as "tooth marks." These range from "pits" (in Binford's [1981b] terminology) to sublinear and linear gouges (see Figure 6.1). These were counted separately from "striations," which I defined as extremely fine (i.e. ≤1 mm wide), and which depressed but did not break the cortical surface. This dichotomous classification in fact obscures the continuous nature of this kind of modification, which would have made ascription of carnivore actors a difficult task if based on this kind of damage alone. I cannot be certain that all cases where cortical lamellae have been removed were in fact carnivore gnaw marks. However, I chose to disregard striations, simply because I could not rule out sediment compaction and trampling as their cause. Where toothmarks occurred near break edges, or in trabecular bone, or in sufficient numbers to warrant the inference of carnivore modification, I felt fairly confident ascribing the term "gnawing" to the traces on a given specimen. These conservative criteria are reflected in the gnawing data presented below.

Haynes (1983b:166) observes that hyenas often leave 10 mm wide toothmarks on cortical and subchondral bone. Wolf tooth marks, on the other hand, are generally far smaller—as wide as 2-3 mm, but commonly less. However, whereas it is possible to identify hyena modification unequivocally from the presence of 1 cm-wide tooth marks, it is impossible to say whether 2-3 mm-wide tooth scratches are the product of the smaller-diameter wolf's teeth or the result of the larger hyena's teeth affecting the bone only slightly. There is no way out of this conundrum, unfortunately, except to appeal to separate lines of evidence, including comparison of the abilities of hyenas and wolves to delete portions of elements, which will be discussed later in this chapter.

Figure 6.1 Adult *Ursus spelaeus* right radius diaphysis showing carnivore tooth marks. Specimen number 105A64. Scale bar length is 1 cm.

Following are the categories of observation made on Pod hradem's assemblage:

1) Binford (1981b) asserts that removal of less than 50% of a bone cylinder through some of its length is unmistakable evidence of canid gnawing (termed "*channelling*"). It is commonly seen, for example, in Dr. Haynes's collection of deer bone gnawed by captive wolves (Haynes 1981).

2) I assessed the relative roughness or smoothness of the break surface, for a character called "*break surface texture.*" This can be helpful in sorting out fractures that occurred when the bone was green, from those that occurred after the bone had lost some of its freshness, or had dried out altogether before breaking.

3) Any circumscribed breach of cortical or subchondral bone was termed a "*puncture*," and was taken as evidence of the effects of carnivore canines or premolars (in the case of hyenas). Any punctures that were 5 mm or more in diameter were taken to be hyena damage, on the model of Stiner (1990a).

4) Conchoidal fracture scars on broken bone surfaces were termed "*flake scars*," and were taken as evidence of gnawing. Sometimes the flakes were large, having been "levered off" the diaphysis, as described by Haynes (1982), and as illustrated in Figure 6.2.

5) Less-than-5 mm-wide flakes removed from the fracture margin are termed "*chipping back*," again after Binford (1981b), who ascribed this type of modification to carnivores.

6) When carnivores gnaw bone they often hold it between their paws and lick one end (e.g. Haynes 1980). During these episodes their rough tongues are capable of rounding broken surfaces. The other end can also be smoothed through contact with the ground. I sought evidence of licking, or friction with fur or sediment, indicative of carnivore behavior. Figure 6.3 illustrates this kind of modification in a specimen whose other end is not similarly affected. I recorded "*marginal rounding*" when it occurred on the broken surfaces, and ascribed such modification to carnivore activity, because it is unlikely to occur for any other reason. When rounding occurred over the entire surface of a specimen, it was recorded simply as "*rounding.*"

7) Rare, unmistakable remodelling of bone by hyena digestive acids was noted (as described by Sutcliffe 1970).

The tooth marks I observed are almost universally less than 5 mm wide, and are thus suggestive of wolf modification (Haynes 1983b:166,167). Indeed, wolves may have been responsible for the majority of damage to the Pod hradem assemblage. Almost all the tooth marks on cortical bone resemble those shown in Figures 6.1 and

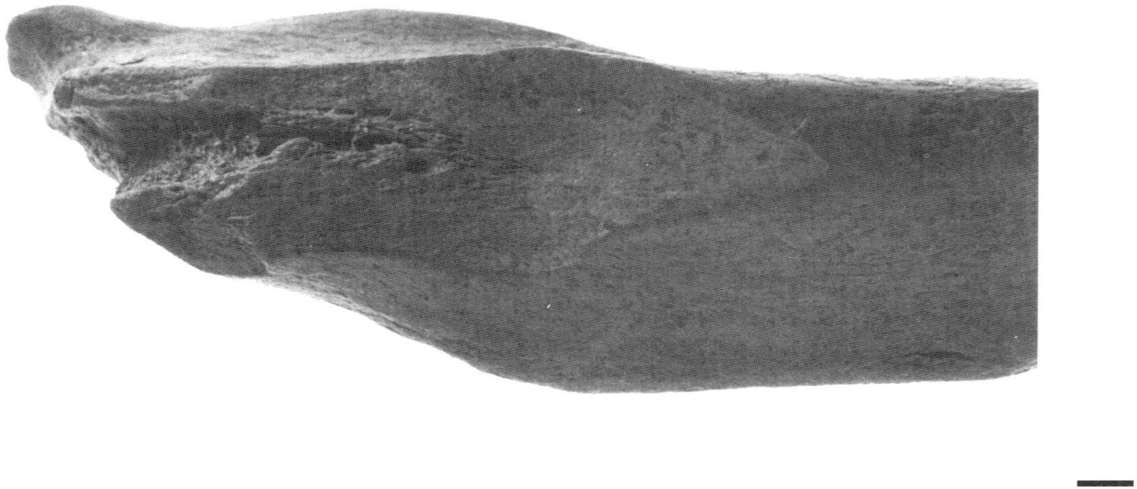

Figure 6.2 Adult *Ursus spelaeus* distal left humerus with large bone flake levered off by carnivore gnawing. Specimen number 127A1. Scale bar length is 1 cm.

Figure 6.3 Subadult *Ursus spelaeus* distal right femur showing marginal rounding, caused by abrasion against a carnivore's tongue or the sedimentary substrate during gnawing. Specimen number 125E54, same specimen as figure 6.4. Scale bar length is 1 cm.

6.4, which look like the wolf modification depicted by Haynes (e.g. 1983a:104, Figure 5).

Table 6.1 presents the cumulative observations of all evidence for carnivore gnawing on the Pod hradem assemblage. All elements have been affected to varying degrees. Percentages of specimens gnawed range from 33% for the small sample of sternal elements to 100% for the innominate, baculum and patella. Most elements have been affected at greater than 50%, suggesting a wolf homesite pattern.

The absence of subadult epiphyses is not simply the result of separation from the metaphyses and subsequent destruction due to fragility. In almost every case, subadult limb bone ends have been chewed off, leaving ragged or crenulated break margins on the diaphysis, crushed compact bone where teeth have been dragged across the break, furrowed cancellous tissue, or tooth scratches near the break, or all of the above (see Figure 6.4). This could only have occurred if the epiphyses, which are attached to the diaphysis by strong metaphyseal tissue, had first been destroyed. By contrast, fetal/neonate epiphyses could not have preserved, even if they were left undisturbed, because they are not ossified at that stage of development.

Discovery of marks on fetal/neonate specimens that appear similar to tooth marks on the larger ones probably stems from the inability of my criteria to distinguish between scratches that can result from trampling and those that are the product of animal gnawing. Especially in the structurally fragile bone of fetal/neonates, much lighter pressure would be required to produce a mark analogous to gnaw marks in adult cortical bone. Thus, where trampling on adult bones might produce very fine striations when sedimentary particles are pressed against the tissue, similar events could easily affect fetal/neonate bone more severely, resulting in deeper scratches. No other kind of damage was observed in conjunction with the few "scratches" on the fetal/neonate sample. For this reason, I cannot think that the marks on these specimens are truly the product of animal gnawing. Indeed, the very presence of fetal/neonate specimens provides evidence that these individuals were, for the most part, left alone once they had died—any large carnivore would simiply have devoured the bones of such small creatures, leaving nothing behind.

Table 6.2 gives the number of occurrences of punctures, flake scars, chipping back, marginal and overall rounding. There were no incidences of what Binford has called channelling, nor of rough break margins that would have implied dry bone breaks. The low numbers of punctures, flake scars on break edges, and chipping back suggests either a low degree of bone modification by competent animals, or a testament to the variability along those dimensions of

Figure 6.4 Subadult *Ursus spelaeus* proximal right femur showing crushing, crenulation, tooth gouges and pits, all evidence of carnivore gnawing. Specimen number 125E54, same specimen as figure 6.3. Scale bar length is 1 cm.

Table 6.1. Pod hradem Cave Stratum 17: carnivore gnawed cave bear skeletal elements (all ages).

	ADULT			SUBADULT			FETAL/NEONATE		
	GNAW	N	%	GNAW	N	%	GNAW	N	%
CRANIUM	23	34	68	14	17	82	57	101	56
MANDIBLE	34	60	57	22	26	85	21	36	58
HYOID	10	18	56	n/a	0	n/a	n/a	0	n/a
VERTEBRAE									
INDET	5	12	42	1	4	25	0	1	0
CERVICAL	29	35	83	4	6	67	n/a	0	n/a
THORACIC	54	84	64	7	9	78	0	2	0
LUMBAR	22	31	71	4	6	67	n/a	0	n/a
SACRAL	3	7	43	n/a	0	n/a	n/a	0	n/a
RIB	148	165	90	15	19	79	2	16	13
STERNUM	2	6	33	n/a	0	n/a	n/a	0	n/a
SCAPULA	18	24	75	6	8	75	0	1	0
HUMERUS									
PROXIMAL	4	5	80	n/a	0	n/a	n/a	n/a	n/a
SHAFT	18	25	72	18	20	90	1	15	7
DISTAL	4	5	80	n/a	0	n/a	n/a	n/a	n/a
RADIUS									
PROXIMAL	4	4	100	1	1	100	n/a	n/a	n/a
SHAFT	12	15	80	27	28	96	1	4	25
DISTAL	1	2	50	1	1	100	n/a	n/a	n/a
ULNA									
PROXIMAL	1	4	25	n/a	0	n/a	n/a	n/a	n/a
SHAFT	7	12	58	20	22	91	3	9	33
DISTAL	1	1	100	n/a	0	n/a	n/a	n/a	n/a
INNOMINATE	5	5	100	2	3	67	9	13	69
OS PENIS	15	15	100	2	2	100	n/a	0	n/a
FEMUR									
PROXIMAL	6	9	67	1	2	50	n/a	n/a	n/a
SHAFT	25	44	57	26	29	90	6	19	32

Table 6.1, continued.

	ADULT			SUBADULT			FETAL/NEONATE		
	GNAW	N	%	GNAW	N	%	GNAW	N	%
DISTAL	5	6	83	1	1	100	n/a	n/a	n/a
PATELLA	11	11	100	2	2	100	0	1	0
TIBIA									
PROXIMAL	5	5	100	n/a	0	n/a	n/a	n/a	n/a
SHAFT	14	19	74	12	15	80	1	3	33
DISTAL	0	7	0	3	3	100	n/a	n/a	n/a
FIBULA									
PROXIMAL	2	2	100	n/a	0	n/a	n/a	n/a	n/a
SHAFT	21	21	100	8	8	100	0	1	0
DISTAL	6	6	100	n/a	0	n/a	n/a	n/a	n/a
METAPODIALS									
PROXIMAL	44	58	76	2	3	67	n/a	n/a	n/a
SHAFT	18	60	30	8	11	73	1	3	33
DISTAL	41	51	80	n/a	0	n/a	n/a	n/a	n/a
PODIALS	31	32	97	5	6	83	n/a	0	n/a
PHALANGES	25	40	63	3	5	60	n/a	0	n/a

Table 6.2. Pod hradem Cave Stratum 17: cave bear with conchoidal flake scars, punctures, chipping back, marginal rounding, and systemic rounding.

	FLAKE SCARS		PUNCTURES		CHIPPING BACK		MARGINAL ROUNDING		OVERALL ROUNDING	
	NUMBER	PERCENT	NUMBER	PERCENT	NUMBER	PERCENT	NUMBER	PERCENT	NUMBER	PERCENT
AFFECTED	26	1.8	25	1.7	47	3.2	720	49	87	6
UNAFFECTED	1431	98.2	1432	98.3	1410	96.8	737	51	1370	94
TOTALS	**1457**	**100**	**1457**	**100**	**1457**	**100**	**1457**	**100**	**1457**	**100**

modification. And, contrary to the norm for hyena utilization, no shaft fragments evince the qualities of "bone shrapnel" described by Sutcliffe (1970), among others. If hyenas had been seriously interested in within-bone nutrients, one would expect to see such modification.

There is, however, unmistakable evidence of occasional hyena utilization, such as the remodelled bone seen in Figure 6.5, but it is extremely rare. If hyenas had been the primary actors affecting this assemblage, such evidence might be expected to be more frequent. Aside from a single 6.5 mm-diameter pit pictured in Figure 6.6, the deeply furrowed trabecular bone of two proximal long bones, and some equivocal mandible damage described later on, there is little evidence of the effects of the hyena's large teeth in the Pod hradem assemblage.

Finally, a word on the possibility that bears had gnawed a portion of the assemblage. Haynes (1983b:168) describes bear modification as "planing" or "shearing," sometimes with parallel tooth furrows in cancellous bone. I noted some shearing on vertebral centra and processes, as well as some planar erosion on several other specimens, but none with the parallel furrows that Haynes observed. Thus, it is reasonable to rule out bears as significant modifiers of this assemblage.

In summary, small toothmarks, high numbers of gnawed juvenile bones, and low numbers of modification uniquely diagnostic of bears, wolves or hyenas, presents a somewhat blurred picture of the actors responsible for gnawing the Pod hradem Stratum 17 cave bears.

Differential Survival of Element Portions

Chapter 5 looked at the Stratum 17 sample using element frequencies and percent survival based on my MNE estimates. I used MNE to compare Pod hradem's data with other published assemblages of known origin, to get an idea of the likely authors of the cave bear sample. Unfortunately, MNE is not very descriptive of the carnivore modification to an assemblage, because it cannot inform about the individual specimens and their condition. Indeed, MNE can even be misleading, if one is interested in knowing exactly what comprises an assemblage. For example, a small portion of an element may be all that is required to denote the presence of that element. Thus, for estimates of MNE, one fragment is equal in value to an entire element. But a fragment tells a different story than a whole element when one is interested in characterizing patterns of modification to ascertain the actor or actors responsible. Thus, in order to allow further comparisons between kinds of modification potentially attributable to either wolves or hyenas, in this section I employ a more descriptive summary of the particulars of modification to individual specimens.

Figure 6.5 Bone fragment remodelled by hyena gastric juices. Specimen number 128E7. Scale bar length is 1 cm.

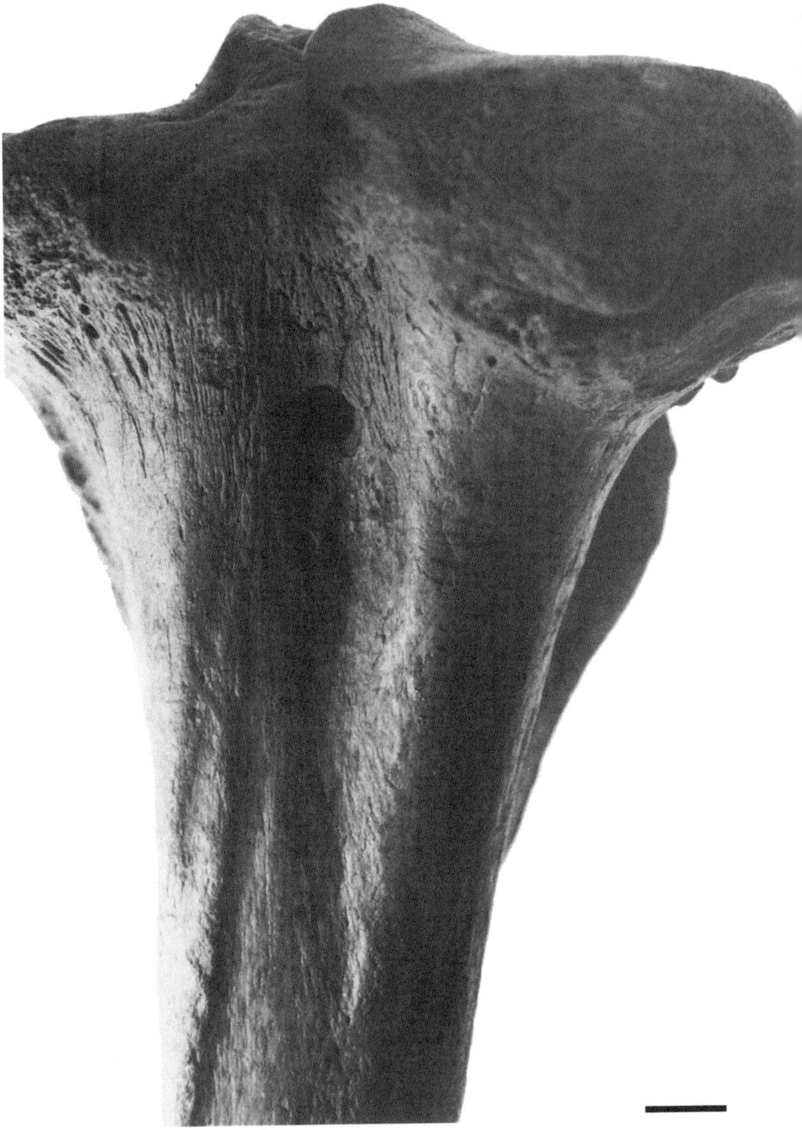

Figure 6.6 Adult *Ursus spelaeus* proximal tibia with probable hyena tooth puncture. Specimen number 105A61. Scale bar length is 1 cm.

Table 6.3 presents a complete inventory of the cave bear specimens in Stratum 17. In contrast to earlier data presentation (Tables 4.5 and 5.1), I have chosen to look at three age categories in this inventory: fetal/neonate, subadult, and adult. A glance at Table 6.3 will confirm that a division according to three age classes reveals much of interest about the structured differences in modification to specimens of different ages. The great range of sizes and developmental stages present in this ursid assemblage means that any species modifying these bones would have had vastly different consequences on carcasses of different developmental ages. Newborn cave bears, for example, were about the size of rats. Thus, they are extremely small when compared with adult specimens (e.g. fetal/neonate long bones are only a few centimeters in length). Among subadults, a great range of sizes occurs, although specimens of each element, regardless of size, looked morphologically similar. All subadult specimens in the collection retained similar amounts of cancellous bone in the medullary cavities of long bones, in contrast to the adults whose medullary cavities were hollow. Thus, the subadults also formed a natural group, and were no doubt treated as such by the modifying carnivores. By adulthood, around four years of age, males in some populations had grown to be larger than today's largest brown bears. Adult long bones, thus, are robust, and most are on the order of half a meter in length.

Keeping in mind the importance of structural differences among the age classes, in deciding what portions of elements to enumerate in Table 6.3, I used knowledge of the architecture of the cave bear skeleton. For example, juvenile cranial elements are unfused, unlike those of adults. Rather than employ a different set of categories for different age groups, I chose to divide the cranium into four regions that transcend developmental differences, and that would have been treated differently by carnivores as they sought sustenance: frontoparietal, basioccipital, temporal-zygomatic, and maxillofacial. Regardless of developmental age, the mandible lends itself naturally to a bipartite division: one, the more robust horizontal ramus; the other, the very brittle ascending ramus. Adult vertebral fragments were most commonly either from the neural arch, or from the centrum, or consisted of units split along the sagittal plane. The same is true of subadult vertebrae. Among archaeofaunal analysts, it is common to present the proximal, shaft, and distal parts of long bones separately, to monitor differential destruction to epiphyses and diaphyses due to the different nutritive content and density of the three parts (e.g. Binford 1981b; Lyman 1984, 1985). Thus, for long bones a fairly detailed division is the result. The remaining elements are treated unitarily. To supplement the detailed inventory in Table 6.3, I have

Table 6.3. Pod hradem Cave Stratum 17: Inventory of adult, subadult, and fetal/neonate cave bear skeletal elements.

	ADULT	SUBADULT	FETAL/ NEONATE
CRANIUM			
COMPLETE	**2**(1)	0	0
FRONTO-PARIETAL	(3)	*see note	(63)
BASI-OCCIPITAL	0	(1)	(7)
TEMPORAL-ZYGOMATIC	(13)	(1)	(16)
MAXILLO-FACIAL	(14)	(15)	(11)
MANDIBLE			
COMPLETE	**13**	**2**	0
ASCENDING RAMUS	(22)	(3)	(7)
HORIZONTAL RAMUS	(23)	(21)	(29)
HYOID	(18)	0	0
ATLAS			
COMPLETE	**8**	0	0
NEURAL ARCH	(1)	0	0
CENTRUM	0	0	0
CENTRUM+ARCH	0	0	0
AXIS			
COMPLETE	**4**	0	0
NEURAL ARCH	(2)	0	0
CENTRUM	0	0	0
CENTRUM+ARCH	(3)	0	0
CERVICAL VERTEBRAE			
COMPLETE	**7**	**4**	0
NEURAL ARCH	(3)	0	0
CENTRUM	(1)	0	0
CENTRUM+ARCH	(5)	(2)	0
THORACIC VERTEBRAE			
COMPLETE	**45**	**6**	**1**
NEURAL ARCH	(19)	(1)	(1)
CENTRUM	(10)	0	0
CENTRUM+ARCH	(11)	(2)	0
LUMBAR VERTEBRAE			
COMPLETE	**22**	**2**	0
NEURAL ARCH	(6)	(4)	0
CENTRUM	(2)	0	0
CENTRUM+ARCH	(1)	0	0
RIBS	**11**(155)	**6**(13)	(16)
STERNUM	0	(6)	0
SCAPULA	**2**(22)	(8)	(1)
HUMERUS			
COMPLETE	**4**	0	0
PROXIMAL END + COMPLETE SHAFT	0	0	0
PROXIMAL END + PROXIMAL + MID SHAFT	0	0	0
PROXIMAL END + PROXIMAL SHAFT	0	0	0
PROXIMAL END ONLY	(1)	0	0
COMPLETE SHAFT	**1**	**5**	**7** (1)
PROXIMAL + MID SHAFT	0	0	0
MID + DISTAL SHAFT	0	0	0

N.B.: numbers in bold refer to complete specimens; parentheses refer to fragmentary specimens

Table 6.3, continued.

	ADULT	SUBADULT	FETAL/ NEONATE
HUMERUS CONTINUED			
PROXIMAL SHAFT	(9)	0	(1)
MID SHAFT	(5)	0	(1)
DISTAL SHAFT	**1** (5)	**10** (5)	**5**
DISTAL END + COMPLETE SHAFT	0	0	0
DISTAL END + MID + DISTAL SHAFT	0	0	0
DISTAL END + DISTAL SHAFT	(1)	0	0
DISTAL END ONLY	0	0	0
RADIUS			
COMPLETE	**1**	0	0
PROXIMAL END + COMPLETE SHAFT	**1**	0	0
PROXIMAL END + PROXIMAL + MID SHAFT	0	0	0
PROXIMAL END + PROXIMAL SHAFT	(1)	0	0
PROXIMAL END ONLY	(1)	**1**	0
COMPLETE SHAFT	0	**14**	**2**
PROXIMAL + MID SHAFT	0	0	0
MID + DISTAL SHAFT	0	0	0
PROXIMAL SHAFT	(5)	**9** (1)	**2**
MID SHAFT	(3)	(2)	0
DISTAL SHAFT	(4)	**2**	0
DISTAL END + COMPLETE SHAFT	0	0	0
DISTAL END + MID + DISTAL SHAFT	0	0	0
DISTAL END + DISTAL SHAFT	0	0	0
DISTAL END ONLY	0	**1**	0
ULNA			
COMPLETE	**1**	0	0
PROXIMAL END + COMPLETE SHAFT	**1**	0	0
PROXIMAL END + PROXIMAL + MID SHAFT	0	0	0
PROXIMAL END + PROXIMAL SHAFT	**2**	0	0
PROXIMAL END ONLY	0	0	0
COMPLETE SHAFT	0	**4**	**3** (1)
PROXIMAL + MID SHAFT	0	0	0
MID + DISTAL SHAFT	0	0	0
PROXIMAL SHAFT	**1** (3)	**13** (5)	**4** (1)
MID SHAFT	(2)	0	0
DISTAL SHAFT	(1)	0	0
DISTAL END + COMPLETE SHAFT	0	0	0
DISTAL END + MID + DISTAL SHAFT	0	0	0
DISTAL END + DISTAL SHAFT	**1**	0	0
DISTAL END ONLY	**1**	0	0
SACRUM	**1**(6)	0	0
INNOMINATE	(5)	(3)	(13)
BACULUM	**9**(6)	(2)	0
FEMUR			
COMPLETE	**4**	0	0
PROXIMAL END + COMPLETE SHAFT	**3**	0	0
PROXIMAL END + PROXIMAL + MID SHAFT	0	0	0

N.B.: numbers in bold refer to complete specimens; parentheses refer to fragmentary specimens

Table 6.3, continued.

	ADULT	SUBADULT	FETAL/ NEONATE
FEMUR CONTINUED			
PROXIMAL END + PROXIMAL SHAFT	**1** (1)	0	0
PROXIMAL END ONLY	0	(2)	0
COMPLETE SHAFT	**4**	**15**	**9** (4)
PROXIMAL + MID SHAFT	0	0	0
MID + DISTAL SHAFT	0	0	0
PROXIMAL SHAFT	(8)	**2** (1)	**3**
MID SHAFT	(9)	(8)	0
DISTAL SHAFT	**1** (12)	**1** (2)	**3**
DISTAL END + COMPLETE SHAFT	**1**	0	0
DISTAL END + MID + DISTAL SHAFT	0	0	0
DISTAL END + DISTAL SHAFT	(1)	0	0
DISTAL END ONLY	0	(1)	0
PATELLA	**11**	**1**(1)	(1)
TIBIA			
COMPLETE	**5**	0	0
PROXIMAL END + COMPLETE SHAFT	0	0	0
PROXIMAL END + PROXIMAL + MID SHAFT	0	0	0
PROXIMAL END + PROXIMAL SHAFT	0	0	0
PROXIMAL END ONLY	0	0	0
COMPLETE SHAFT	**1**	**9**	**3**
PROXIMAL + MID SHAFT	0	0	0
MID + DISTAL SHAFT	0	0	0
PROXIMAL SHAFT	(8)	**1**	0
MID SHAFT	(3)	0	0
DISTAL SHAFT	0	**4** (1)	0
DISTAL END + COMPLETE SHAFT	0	0	0
DISTAL END + MID + DISTAL SHAFT	0	0	0
DISTAL END + DISTAL SHAFT	**2**	0	0
DISTAL END ONLY	0	**2** (1)	0
FIBULA			
COMPLETE	0	0	0
PROXIMAL END + COMPLETE SHAFT	0	0	0
PROXIMAL END + PROXIMAL + MID SHAFT	0	0	0
PROXIMAL END + PROXIMAL SHAFT	**1**	0	0
PROXIMAL END ONLY	(1)	0	0
COMPLETE SHAFT	**1**	**4**	**1**
PROXIMAL + MID SHAFT	0	0	0
MID + DISTAL SHAFT	0	0	0
PROXIMAL SHAFT	**3**	**1**	0
MID SHAFT	(2)	(2)	0
DISTAL SHAFT	**9**	**1**	0
DISTAL END + COMPLETE SHAFT	0	0	0
DISTAL END + MID + DISTAL SHAFT	0	0	0
DISTAL END + DISTAL SHAFT	**6**	0	0
DISTAL END ONLY	0	0	0

N.B.: numbers in bold refer to complete specimens; parentheses refer to fragmentary specimens

Table 6.3, continued.

	ADULT	SUBADULT	FETAL/ NEONATE
METAPODIALS			
COMPLETE	**51**	0	0
PROXIMAL END + COMPLETE SHAFT	**7**	**3**	0
PROXIMAL END + PROXIMAL + MID SHAFT	0	0	0
PROXIMAL END + PROXIMAL SHAFT	0	0	0
PROXIMAL END ONLY	0	0	0
COMPLETE SHAFT	**2**	**8**	**3**
PROXIMAL + MID SHAFT	0	0	0
MID + DISTAL SHAFT	0	0	0
PROXIMAL SHAFT	0	0	0
MID SHAFT	0	0	0
DISTAL SHAFT	0	0	0
DISTAL END + COMPLETE SHAFT	0	0	0
DISTAL END + MID + DISTAL SHAFT	0	0	0
DISTAL END + DISTAL SHAFT	0	0	0
DISTAL END ONLY	0	0	0
PODIALS	**32**	**6**	0
PHALANGES	**38**(2)	**1**(4)	0

* There are, indeed, sub-adult fronto-parietal specimens. However, the author was unable to identify them with certainty to taxon. It is probable that 104 parietal fragments, identified as large mammal, are sub-adult cave bear. In addition, due to data recording inconsistency in the Budišov field laboratory, some cave bear sub-adult frontal specimens were recorded as fetal/neonate. Thus, there are probably at least 110 specimens in the category "sub-adult cave bear fronto-parietal."

N.B.: numbers in bold refer to complete specimens; parentheses refer to fragmentary specimens

used its data as the basis for presentation of the more-refined NISP and MNE figures given in Table 6.4. Long-bone data are thus presented in a form more conducive to interassemblage comparisons than was the case in Tables 4.5 or 5.1.

Differential survival of elements across ages was documented in Chapter 5. Tables 6.3 and 6.4 reveal that certain *parts* of elements of the cave bear skeleton fared better than others in Pod hradem's Stratum 17, especially when the different age groups are compared. One is struck by some obvious differences between adults and subadults. First, there are always complete adult specimens; there are almost never complete subadult pieces. Compared to the adults, many more juvenile cranial parts survive. Vastly more adult vertebral parts are represented. There are many more shaft cylinders among subadults than in the adult sample. Finally, there are relatively few adult shaft fragments. A detailed examination follows.

Cranium. The greater part of 13 adult crania in the assemblage appears to have been deleted by carnivore modification. The basicranial bones have fared poorly. Only portions of 3 basicrania survive, 2 of which belong to complete crania. And, with the exception of some portions of the premaxilla/maxilla, very few facial fragments have preserved. In addition, nearly 70% of adult cranial specimens are tooth-marked. Marks on and near break surfaces of the maxillae, and the regular nature of the damage to the adult maxillary portions visible in Figure 6.7 attest to carnivore modification of this element. One possible explanation for the absence of facial and basicranial portions can be found in the habitual behavior of carnivores, which regularly access the nutritious brain tissue through the fragile bones of the face and basicranium (Binford 1981b; Blumenschine 1986; Brain 1981:72; Haynes 1980).

In contrast to the adult sample, no complete subadult crania survive. As was the case with adults, subadult basicrania have not survived, and only one fragment of the temporal is present. Only a few fragments of the more durable portions of the subadult maxilla have survived. Whereas 6 of 13 adult maxillae survive, only 6 of 26 juvenile maxillae have preserved. It should be noted that, although Table 6.3 records no subadult frontoparietal portions, it is probable that a large number preserved from this age class, as I suggested previously. Due to the conservative criteria employed in taxonomic identification, and the difficulty of making secure ascriptions in the field, it is likely that most of the 104 (NISP) cranial fragments identified as large mammal and mentioned in Chapter 4 are indeed those of cave bear subadults. Not only are these cranial portions fragmentary, but also gnaw marks are common on the parietal fragments, and several of the frontal bones

Table 6.4. Pod hradem Cave Stratum 17: Inventory of adult, sub-adult, and fetal/neonate cave bear skeletal elements.

	ADULT		SUBADULT		FETAL/NEONATE	
	NISP	MNE	NISP	MNE	NISP	MNE
CRANIUM	39	6	17	8	101	19
½ MANDIBLE	63	17	26	10	36	23
HYOID	18	2	0	0	0	0
VERTEBRAE						
INDET.	12	2	4	1	1	1
CERVICAL	35	35	6	6	0	0
THORACIC	84	78	9	8	2	2
LUMBAR	31	34	6	3	0	0
SACRUM	7	2	0	0	0	0
RIB	178	151	19	4	16	3
STERNUM	6	2	0	0	0	0
SCAPULA	24	9	8	3	1	1
HUMERUS						
OVERALL	29	9	20	16	15	12
PROXIMAL	5	5	0	0	n/a	n/a
SHAFT	26	6	20	16	15	12
DISTAL	5	5	0	0	n/a	n/a
RADIUS						
OVERALL	16	5	30	24	4	4
PROXIMAL	4	4	1	1	n/a	n/a
SHAFT	15	3	29	24	4	4
DISTAL	1	1	1	1	n/a	n/a
ULNA						
OVERALL	13	10	22	18	9	9
PROXIMAL	4	4	0	0	n/a	n/a
SHAFT	11	5	22	18	9	9
DISTAL	3	3	0	0	n/a	n/a
INNOMINATE	5	5	3	1	13	11
BACULUM	15	12	2	2	0	0
FEMUR						
OVERALL	45	11	32	20	19	16
PROXIMAL	9	9	2	2	n/a	n/a
SHAFT	45	13	29	19	19	16
DISTAL	6	5	1	1	n/a	n/a
PATELLA	12	11	2	2	1	1
TIBIA						
OVERALL	19	11	18	14	3	3
PROXIMAL	5	5	0	0	n/a	n/a

Table 6.4, continued.

	ADULT		SUBADULT		FETAL/NEONATE	
	NISP	MNE	NISP	MNE	NISP	MNE
Shaft	19	6	15	14	3	3
Distal	7	7	3	3	n/a	n/a
Fibula						
Overall	23	10	8	5	1	1
Proximal	2	2	0	0	n/a	n/a
Shaft	23	6	8	5	1	1
Distal	6	6	0	0	n/a	n/a
Metapodials						
Overall	61	60	11	11	3	3
Proximal	58	60	3	3	n/a	n/a
Shaft	60	60	11	11	3	3
Distal	51	51	0	0	n/a	n/a
Podials	32	32	6	6	0	0
Phalanges	40	40	5	5	0	0

N.B.: These numbers do not reflect MNE for complete elements as calculated in the field laboratory.

Figure 6.7 Adult *Ursus spelaeus* facial portions showing range of carnivore modification. Specimen numbers, from left to right, upper: 109Ca55, 114a1; lower: 118E11, 127D52, 106A4. Scale bar length is 1 cm.

show a kind of repetitive gnawing damage to the supraorbital eminence, whereby the bone has been scraped off, leaving the frontal sinus exposed.

In a general way, differential susceptibility to destruction could explain the much lower proportional representation of portions of the subadult and fetal/neonate cranium. The subadult face, being smaller and more easily fragmented due to its developmental immaturity, would have been more susceptible to destruction than the adult (e.g. Behrensmeyer and Dechant-Boaz 1980:91; Haynes 1980; Richardson 1980). More specifically, three major destructive agents (discussed below) could have been responsible for the modification to the subadult cranium, and ultimately to the deletion of the face and base.

First, hyenas would likely have impacted the skeletons of juveniles more severely than those of adults, probably resulting in complete destruction of the juvenile cranium (Blumenschine 1986). Blumenschine (1986:647) has observed that, because of the lack of desirable marrow in juvenile long bones, hyenas regularly access their head contents, for the nutritious brain tissue they contained, after the hindlimb and before the forelimb is consumed (i.e. early in the sequence). Thus, a hyena would not likely have spared the cranial bones in that event; they would have been devoured along with the flesh they contained. The presence of high numbers of what are probably juvenile cranial bones, then, may preclude hyenas as the primary modifiers of (at least) this portion of the assemblage.

Second, wolves are capable of consuming the crania of large animals when they heavily utilize a carcass (e.g. 150 kg subadult bison [Haynes 1982]). Thus they cannot be ruled out as the authors of the modification to the subadult portion of the Pod hradem assemblage.

Third, as I have previously mentioned with regard to subadult and fetal/neonate bones in general, there is a chance that, even if the crania of subadults had never been attacked, they would still have succumbed to the rigors of trampling. Comminution resulting from such forces might have rendered some pieces too small to recognize given the excavation techniques employed.

Mandible. Although the mandible is the most frequently occurring element in the adult cave bear sample from Stratum 17, and has thus been used to estimate MNI for the entire adult sample, it has nevertheless suffered significant deletions—only 17 of the 26 expected hemimandibles survive in any form. Adult mandibles record few gnaw marks, although there is clear evidence of a carnivore pattern of modification in one specimen (Figure 6.8), that, given the robusticity of the cave bear mandible, was likely the result of hyena gnawing.

Relative survival of the mandible conforms to a model of differential susceptibility to destruction by carnivores. In any environment destructive to bone, the densest bones are those most likely to survive (Brain 1981:19-21; Lyman 1984, 1985; Grayson 1988, 1989). The horizontal ramus of the adult mandible is very durable, with 10 mm of dense cortical bone not uncommon. The ascending ramus, on the other hand, is quite thin and therefore relatively fragile, being only a few millimeters thick, and very broad and flat, for the attachment of the cave bear's enormous temporalis muscle. This would explain higher survival rates of the more durable horizontal ramus. Subadult and fetal/neonate mandibles are relatively more fragile, and could have been easily consumed by wolves and hyenas alike. Moreover, once broken into small enough pieces, either by trampling or chewing, the mandibles of younger individuals could easily have been missed by even the most fastidious excavators.

The mandible also contains some edible and fat-rich tissue in the posterior horizontal ramus, which is also associated with the very large masseter muscle. Together, these characteristics of the cave bear mandible must have made the posterior portion a target for consumption (cf. Binford 1981b:66). Wolves and dogs, as well as hyenas, tend to gnaw at the posterior and lower portions of mandibles, to extract the nerve tissue and fat beneath the teeth, and leave the rest of the element, if it cannot be broken further or swallowed whole (e.g. Brain 1981:70). Because it is unlikely that adult mandibles would have been devoured in their entirety, the absence of 9 adult hemimandibles from what would be expected suggests that they may have been transported out of the cave, or alternatively, that they occur in the unexcavated part of Stratum 17.

Atlas. Of the 13 adults represented in this assemblage, 9 atlases have survived (a 69% survival rate). However, even these durable elements have undergone significant modification from the natural state. Because most of the specimens have undergone reductions, total destruction cannot be ruled out as the explanation for the "missing" atlases, even though it is a very durable bone in the adult.

Absence of juveniles can be understood partly in terms of the fragility, small size, and developmental immaturity of fetal/neonate vertebrae, and partly due to the proximity of these elements to the head of the subadult. Carnivores must remove the fragile bones at the base of the skull in order to get at the brain tissue. In so doing they are likely to impact the atlas. It would, therefore, have been devoured along with the bones of the basicranium. Although a plausible

Figure 6.8 Adult *Ursus spelaeus* right hemimandible showing probable hyena modification. Note rounding on broken surface. Specimen number 107E4. Scale bar length is 1 cm.

explanation for their absence, carnivore modification seems equally as likely as destruction by other mechanical processes or size-related recovery bias. Because of their small size, it is somewhat improbable that the unfused centers of ossification of fetal/neonate vertebrae would have been recovered as isolated specimens by the excavators. Moreover, such small and fragile elements would have been most susceptible to destruction through trampling. Subadult specimens, being somewhat larger, might have been collected more consistently, however, their developmental immaturity would still have rendered them susceptible to the jaws of carnivores or comminution through trampling. Therefore it is plausible that juvenile atlases were deleted by a combination of destructive processes, and not necessarily by carnivores.

Axis. Like the atlas, the adult axis has a high survival percentage: 8 of the thirteen expected (or 62%). The body and cranial articular surface are most commonly preserved intact and are very resistant to fragmentation. Conclusions regarding the axis are similar to that of the atlas. Adult elements are more durable. Subadult bones would be more susceptible to destruction than adult specimens. Fetal/neonate bones are susceptible to destruction as well as omission due to the recovery techniques.

Other Cervical Vertebrae. Of the 65 other cervical vertebrae expected from 13 individuals, 18 have survived—only a 28% survival rate. The centrum is most commonly preserved. The articular surfaces and spinous processes are less well preserved—survival relative to that of the centrum ranges from 50% to 90%. Relative to centrum survival, the transverse processes survive only about 40% of the time. Gnawing damage appears most frequently on processes, with the prominent transverse processes suffering most of the damage.

Only 1% of the expected juvenile cervical vertebrae are represented. The presence of those 1% argues that the same processes contributing to the absence of juvenile atlases and axes were no doubt in operation with these other cervical elements.

Actualistic studies of extant carnivores shed some light on the differential survival of vertebral portions. Adult vertebral processes are the most commonly affected part of the vertebrae when wolves utilize the carcass of a large animal (Binford 1981b:66), with sometimes entire processes removed (Haynes 1980:344). Vertebrae gnawed by hyenas, on the other hand, tend to sustain substantial damage also to the body of the vertebra (Brain 1981:72). At Pod hradem vertebral centra have the highest survival rates, with only rare gnawing observed. Thus, with very little reduction of the bodies of cervical vertebrae, it seems probable that wolves were the ones doing the most damage.

Thoracic Vertebrae. With 78 (MNE) elements recovered, adult thoracic vertebrae survived at 43% of what would be expected from 13 individuals. As was the case with cervical vertebrae, the centrum survives most often. Transverse processes survive only about half as often. Spinous processes survive in the lowest frequencies of all. Gnawing damage is responsible for deletion of the processes. Tooth marks are more common on thoracic vertebra than cervical, and all are less than 3 mm wide.

Although the more-dense portions of the neural arch survive more frequently than the spongy centrum in juvenile specimens, the overall small size and fragility of subadult and fetal/neonate specimens, in the face of several destructive processes, probably contributed to their virtual absence. Adult cave bear vertebrae are, on the other hand, large bones with many protuberances in association with some fairly meaty and desirable portions of the carcass (especially the entrails which the ribs encase). Thus, it is no surprise that, though large and spiky, they should be deleted by destructive carnivores, and that the long spinous process should have suffered the greatest damage. Once again, the small size of tooth marks, coupled with the overall pattern of destruction on the processes, suggests that wolves were the most likely modifiers of the adult bones.

Lumbar Vertebrae. Almost 44% of the adult lumbar vertebrae have survived (MNE=34). Centrum, spinous process and articular surfaces almost always survive at the same rate. Transverse processes are less well represented, surviving at less than 15% relative to that of the centrum. Modification to adult specimens includes removal of substantial bone mass from the transverse processes, followed by rounding of the broken surfaces, resulting in the processes having the appearance of having been ground to a plane surface (see Figure 6.9).

With the exception of the curious rounding on the transverse processes, lumbar vertebrae are affected in much the same way as were the cervical and thoracic vertebrae, above. Their processes have been deleted, and the bodies have been left more or less intact. Gnawing damage includes small tooth gouges, dents, and the removal of processes. Hyenas might have left more tell-tale large tooth marks. Thus, once again it seems less likely that hyenas were involved, and more probable that wolves modified the majority of these specimens.

Ribs. At 42%, adult cave bear ribs have survived at more or less the same rate as the vertebrae, with many more fragments of the undiagnostic shaft surviving than any other zone. The angle and the tubercular zones are most commonly preserved intact, while the head

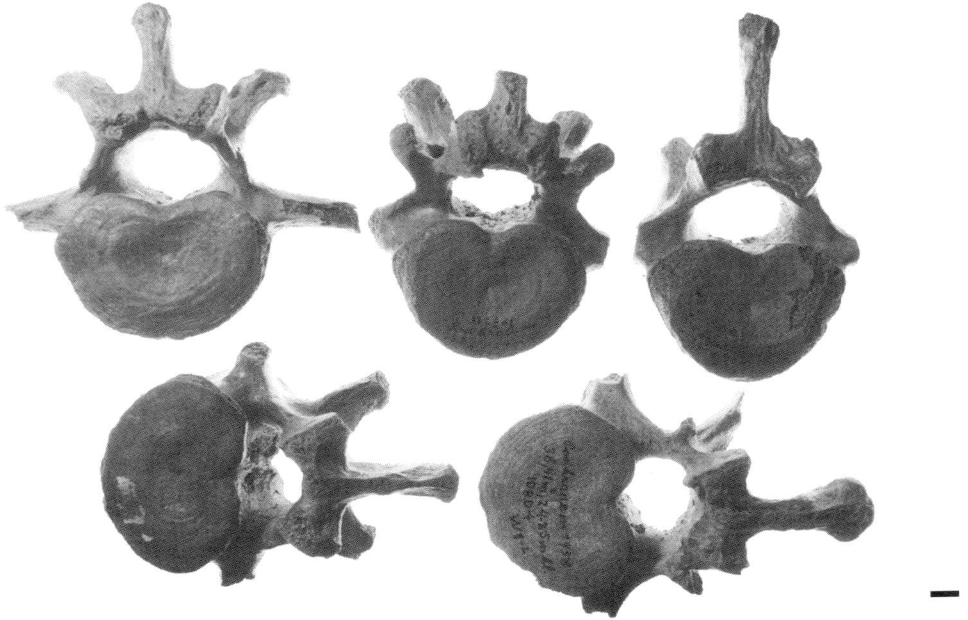

Figure 6.9 Adult *Ursus spelaeus* lumbar vertebrae showing range of carnivore modification, including "shearing" on transverse processes. Specimen numbers, clockwise from upper left: 114B50, 122C63, 120A3, 131C50, 108D4. Scale bar length is 1 cm.

survives intact only about half as often. Gnawing damage is most frequent on the head and angle, and is almost always of the kind illustrated in Figures 6.10 and 6.11, i.e. small tooth drags, and removal of the superior angular margin.

Such minimal damage to the surviving adult ribs, including less than 5 mm-wide tooth drags, suggests, once again, that hyenas were not the primary modifiers of this assemblage. As with the axial elements, the small size and fragility of the juvenile rib makes its destruction by any number of processes more likely than that for the adult element, resulting in its virtual absence from the assemblage.

Scapula. Only 9 (MNE) of an expected 26 adult scapulae survive (or 35%). Two of these are almost complete. Of the fragmentary specimens, the dense glenoid fossa occurs intact most frequently, with portions of the blade and spine surviving intact less than half as often. Fragments of the blade are most frequent, but not numerous enough to account for the missing scapulae. Repetitive damage to the coracoid process is observed.

Wolves commonly remove the acromion, coracoid, and scapular blade, leaving only the glenoid portion (Binford 1981b:66). In a dog yard, Haynes (1980:345) found scapulae little used. Furthermore, hyenas usually leave the scapula untouched, perhaps due to the ease of removal of the flesh. Thus an ambiguous interpretive problem exists with regard to this element. Perhaps the large deltoid muscle and large, spongy acromion process made these scapulae desirable targets.

To explain the differential survival of adult and juvenile scapulae, the size and durability of younger specimens must again be invoked.

Humerus. Of the 26 adult humerus specimens that should exist based on an MNI of 13, 9 have survived (35%), 4 of which are complete. Between 70 and 80% of adult specimens record gnawing damage, although most occurs on proximal and distal epiphyses. With few exceptions, tooth marks are small and shallow. There is tentative evidence of hyena modification, in several wide, deep gouges in one humeral head. However, this speaks of extremely light utilization by hyenas, since not all specimens were so affected. In contrast to what might be expected from hyena-utilized carcasses, there are remarkably few shaft fragments. This suggests that, on occasion, the medullary cavities of these elements were breached, but this was by no means common.

Destruction of subadult epiphyses and long-bone ends is the most salient feature of the long bones in the Pod hradem Cave assemblage, and humeri are no exception (see Figure 6.12). Of the approximately 16 subadult specimens, only one retains any part of an epiphysis. There are 10 more or less complete shafts, and substantial portions of

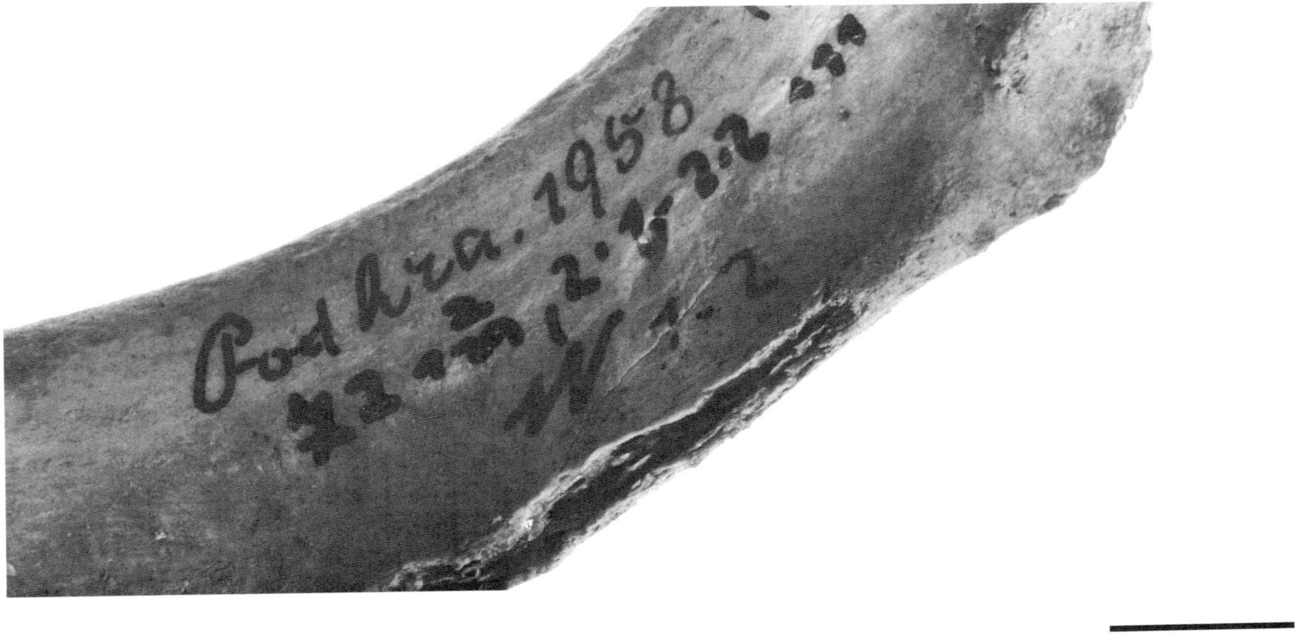

Figure 6.10. Adult *Ursus spelaeus* rib, angle portion, showing typical modification to superior eminence. Specimen number 126A7, same specimen as in figure 6.11. Scale bar length is 1 cm.

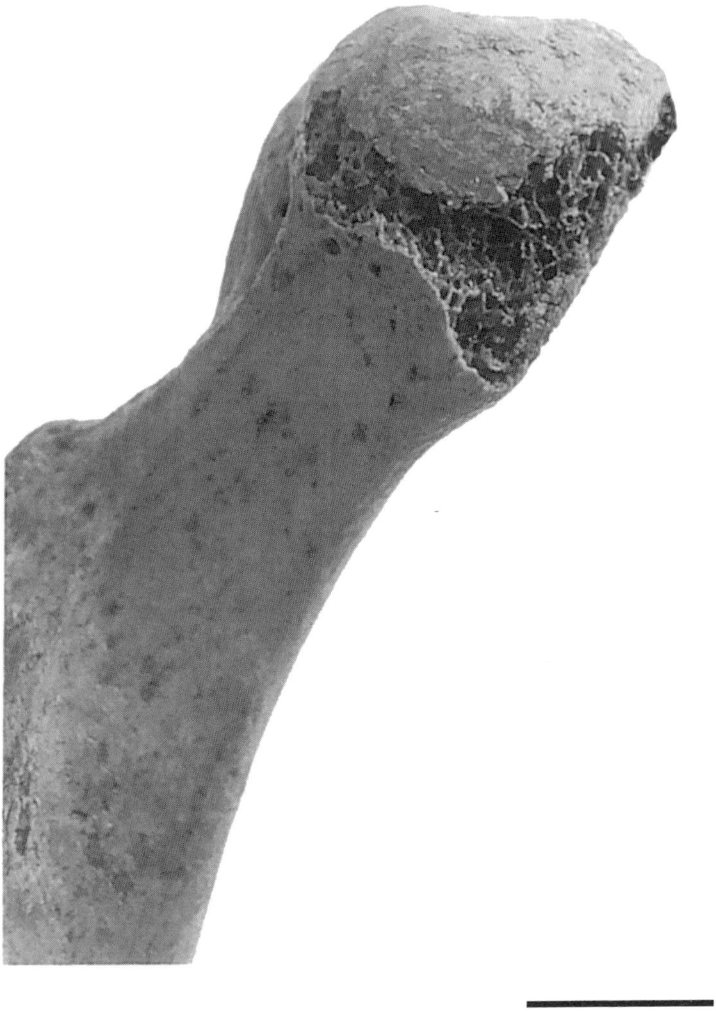

Figure 6.11 Adult *Ursus spelaeus* rib, head portion, showing tooth gouge through trabecular bone. Specimen number 126A7, same specimen as in figure 6.10. Scale bar length is 1 cm.

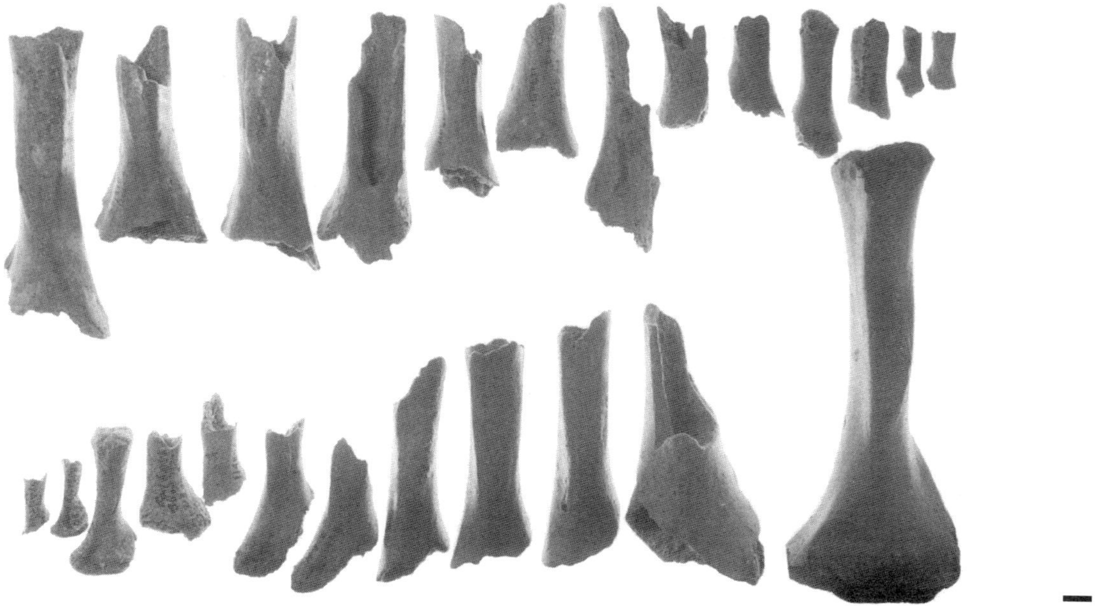

Figure 6.12. Juvenile *Ursus spelaeus* right (upper row) and left (lower row) humeri showing range of modification. Specimen numbers from left to right, upper: 123B7, 125Ea5, 131A51, 129D50, 120b1,127a51, 106Ea1, 110C54, 128B71, 113a54, 129a3, 123C13, 106c29; lower: 118E7, 128E15, 122E7, 107D10, 114E52, 124B2, 124A8, 119D13, 124A10, 125B7, 127D53, 126E6). Scale bar length is 1 cm.

5 mid shafts. Distal diaphysis most commonly occurs, with mid diaphysis a close second. More than 90% of these specimens have been gnawed.

Wolves can remove the proximal humerus of adult caribou (Binford 1981b:66) and bison in heavily-utilized assemblages, such as homesites (Haynes 1980:345). Of the two, bison is closer in size to the adult cave bears. In wolf homesites, humeri are fractured first from the proximal end (Haynes 1982:269). The missing epiphyses in the Pod hradem sample, which must have existed or there could be no shaft fragments, could either be elsewhere in the unexcavated portion of the cave, or have been consumed. But if the latter, and if the remaining shaft fragments were also consumed, why would animals capable of destroying adult long bones leave the shafts of the majority of subadult specimens? Some few fragments of bones remodelled by (probably) hyena gastric acids do occur in the assemblage (Figure 6.5), and some evidence for hyena modification of distal humerus is in evidence, including removal of a large flake levered off the distal shaft of one specimen (Figure 6.2). Thus, it is clear that these animals were present at times. However, the occurrence of many complete cave bear long bones and numerous shafts does argue that, if hyenas were the only species present, they did not consume the bears' bones at all times. The dearth of shaft fragments lends support to this conclusion.

Radius. Of the 26 adult radius specimens that should be present, based on an MNI of 13, just 5 (19%) have survived, of which only 1 is complete. Another is nearly complete, missing only the distal epiphysis. Significantly, the distal epiphysis of the complete specimen is the only one that has survived intact. Of the proximal adult radii, 100% have been gnawed, 80% of the shafts have been gnawed, and 50% of the distal epiphyses have been gnawed. Perhaps because the distal radius contains more spongy bone, its absence in all but one of the 5 adult specimens is understandable.

There are approximately 24 subadult radii, of which 14 are represented by complete shafts (see Figure 6.13). In addition there are 9 proximal shaft portions. Once again, epiphyses are all but absent from this sub-assemblage. All parts of these specimens have been gnawed between 90 and 100% of the time.

Ulna. As with the other paired elements of the adult cave bear at Pod hradem, there should be 26 ulnae. However, just 10 have survived (38%), of which 1 is complete. Another is missing only the distal epiphysis. Another 2 lack the mid shaft and distal end. Proximal epiphyses fare less well in ulnae than in the other long bones. Gnawing occurs on 25% of the proximal epiphyses, 58% of the shafts, and 100% of the distal epiphyses.

Figure 6.13. Juvenile *Ursus spelaeus* left radii showing range of modification. Specimen numbers from left to right: 123D13, 114a7, 108C13, 128A10, 114A2, 118c50, 131A54, 120C53, 126d18, 121B4, 113C51, 114A50. Scale bar length is 1 cm.

No epiphyses survive from the approximately 18 subadult specimens of ulna (see Figure 6.14). Proximal shaft survives more frequently than distal, although this could easily have occurred due to chance. Gnaw marks occur on 91% of the subadult specimens.

Proximal adult epiphyses survive less frequently than those of other long bones. This could be due to the presence of unusual amounts of spongy bone in the olecranon process. Binford (1981b), Haynes (1980:346) and Miller (1969:20) list this part of ulna most likely to succumb in light wolf utilization. Once again, the subadults do not retain either of their epiphyses. And the fetal/neonate specimens are often intact.

Innominate. In the 5 surviving adult innominate specimens (19%), only the acetabular portions have survived (see Figures 6.15 and 6.16). Large portions of the ilium, ischium and pubis have been deleted, with scooping out of the cancellous tissue in evidence in most cases. The pubic zone is completely deleted 60% of the time. All of them are toothmarked.

As previously mentioned, the cave bear innominate is built more like that of humans than that of cervids or bovids. As a result, the substantial spongy portions of the three constituent bones were apparently attractive to those animals entering the carcasses from the anal region, which is the most common point of entry in large carcasses (Binford 1981b:66). Compared with the wolf-killed deer, cave bear innominates survive very rarely. Given that wolves attack the spongy portions of bison and moose innominates (Haynes 1980:347), and since most of the bear innominate is spongy, it should be expected that these carnivores would attack the spongy portions of the cave bear innominate, leaving only the acetabular portion intact.

Sacrum. There was only 1 complete sacrum in the assemblage. In addition to its "Swiss cheese" architecture, which makes it more susceptible to destruction, its location near the anus, the common point of entry, and between the two desirable innominates, can explain its virtual absence.

Femur. Of the 26 adult femora that should be expected, the remains of 11 (42%) have survived. Of the 11, 4 are complete, 3 lack the distal extremity, 1 lacks the proximal extremity, 1 lacks the mid shaft and distal end. There are also 4 complete shafts minus epiphyses. Whole and fragmentary epiphyses survive only 20 to 30 percent of the time relative to the highest surviving portion, the distal shaft. Gnawing is in evidence on 67% of the proximal femur, 57% of the mid shaft, and 83% of the distal end. Subadult femoral survival is greatest in the shaft. Epiphyses survive less than 10% of the time (see

Figure 6.14. Juvenile *Ursus spelaeus* right ulnae showing range of modification. Specimen numbers from left to right: 128A3, 126E7, 120B57, 107c1, 108b1, 105A53, 115C7, 114D54, 107C55, 118E6. Scale bar length is 1 cm.

Figure 6.15. Adult *Ursus spelaeus* left innominate showing typical carnivore modification to ilium. Specimen number 105A3, same specimen as in figure 6.16. Scale bar length is 1 cm.

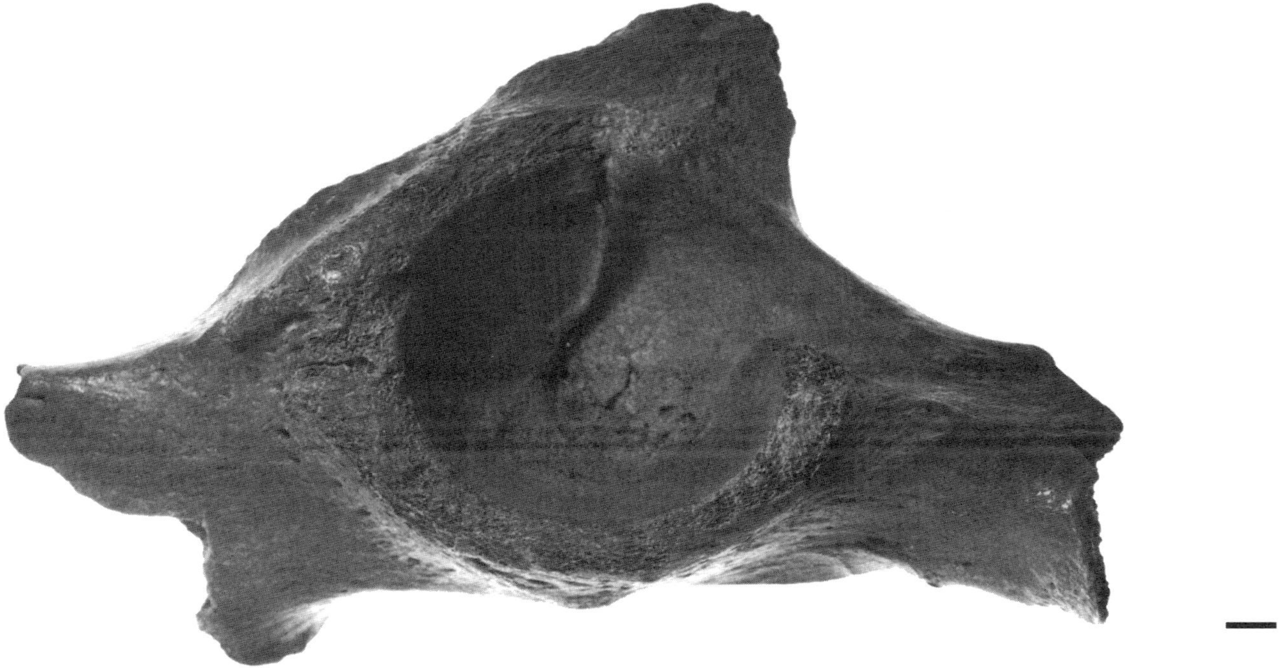

Figure 6.16. Adult *Ursus spelaeus* left innominate showing typical carnivore modification. Specimen number 105A3, same specimen as in figure 6.15. Scale bar length is 1 cm.

Figure 6.17). Gnawing is in evidence on 50% of the proximal ends, 90% of the mid shafts, and all of the distal ends.

Haynes (1982) reports that in heavy utilization of a bison carcass, the femoral condyles are removed before the proximal end. At Pod hradem, femora with distal ends missing outnumber those with proximal ends missing by 3 to 1. Moreover, gnawing is more prevalent distally, in accord with the habits of wolves. In contrast to the more-complete adult specimens, nearly all of the subadult epiphyses have been removed. However, even the quite-small subadult elements have undergone relatively light utilization. This continues the pattern for long bones seen in the forelimb.

Patella. Of the 26 adult patellae, 11 have survived (42%). These are commonly heavily chewed (see Figure 6.18), with numerous wolf-sized pits and furrowed cancellous bone where the cortex has been removed.

The numerous gnawing marks on the adult patella suggest a reason for the virtual absence of subadult specimens: something about them seems to have attracted gnawing behavior. Perhaps it was their position amid the tendons of several large muscles (Gifford-Gonzalez, personal communication). Although the adult bone is very dense and durable, the subadult bone could not have withstood similar pressures, and would have been destroyed at a greater rate.

Tibia. Of the 13 paired tibia to be expected in Stratum 17, 11 have survived (42%), of which 5 are complete, 2 are minus the proximal end, and 1 lacks both epiphyses. Distal ends and fragments outnumber those of the proximal end by a ratio of 5:3. Gnawing is in evidence on 100% of the proximal specimens, 74% of the mid shaft, and on none of the distal pieces.

There are 9 complete juvenile tibial shafts minus epiphyses, three distal epiphyses, and 5 distal shaft fragments. The distal portions survive far better than those of the proximal end (see Figure 6.19). Gnawing is in evidence on 80% of the shaft specimens, and 100% of the distal portions.

Proximal tibia, being predominantly trabecular and subchondral bone and full of muscle attachments, is virtually missing from the adult and subadult sample. Of the adult specimens, 50% are missing the proximal ends, and most are shaft cylinders. As with other documented assemblages (e.g. Binford 1981b:76; Haynes 1980:347), distal tibia almost always survives, whether modified by wolves or hyenas. Numerous cylinders in the subadult sample attest to carnivore modification, with the most-likely actor being the wolf. Wolves channel and chip back the tibia shaft from the proximal end (Binford

Figure 6.17. Juvenile *Ursus spelaeus* femora showing range of modification. Specimen numbers from left to right: 128D66, 126C9, 126d4, 122E1, 125Ea3, 123B50, 108E7, 110C4, 125E54, 108C51, 106C51. Scale bar length is 1 cm.

Figure 6.18 Adult *Ursus spelaeus* patella showing typical modification—wolf-sized tooth pits. Specimen number 123D51. Scale bar length is 1 cm.

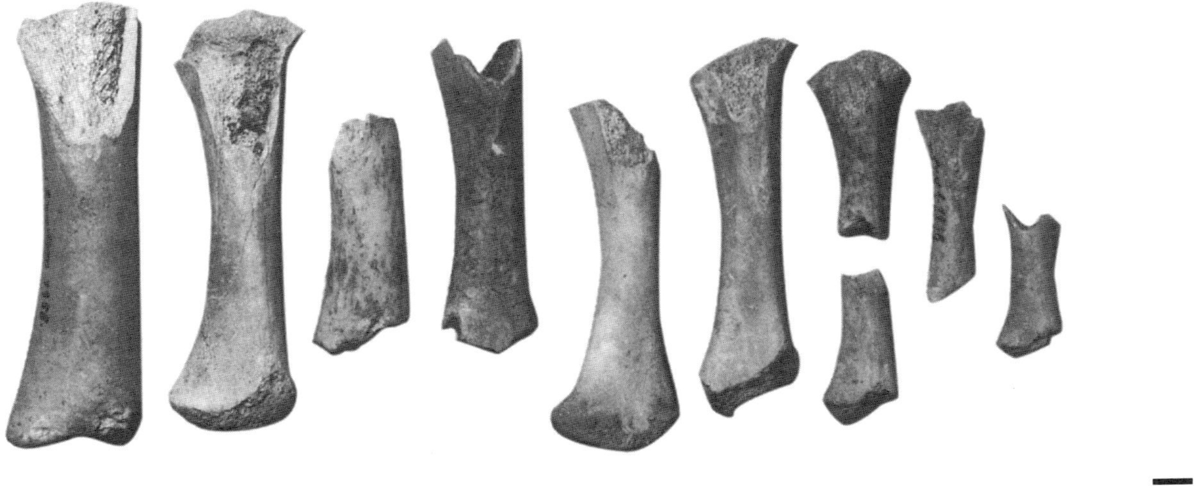

Figure 6.19. Juvenile *Ursus spelaeus* right tibiae showing range of modification. Specimen numbers from left to right: 126c1, 113C50, 125A2, 110A1, 125C52, 106e1, 121C50, 109c50, 110Eb52, 128D52. Scale in cm.

1981b:76; Haynes 1982:269), and are capable of removing the proximal end of animals as large as bison (Haynes 1980:347).

Podials. Only 16 of the more than 200 small bones of the adult manus and pes have survived (i.e. less than 10%). Of these, 97% are gnawed. As discussed previously, wolves and hyenas are capable of swallowing podials complete (Binford 1981b:74). Their low survival rate is probably attributable to the presence of fat-rich pads on the extremities of the plantigrade bears. Any small bones would most likely have been consumed along with the soft tissues.

Metapodials. Metapodials might have suffered the same fate as the smaller bones of the manus and pes. Well within the capabilities of wolves, these would have been swallowed along with the other bones of the limb extremities.

Phalanges. About 25% of the more-than-200 expected adult phalanges have survived. All portions survive in more-or-less-equal proportions, with the distal epiphyses slightly under-represented. Gnawing occurs on proximal and distal epiphyses, and is usually restricted to the removal of trochlea and small (i.e. less than 5 mm) patches of subchondral bone. About 80% of the specimens have been affected in this way. Gnawing appears less frequently on the shafts.

Baculum. There are 12 bacula in this collection. All have been heavily gnawed, with wolf-sized gouges and dents all over the surface. Hyenas would surely have devoured bacula completely along with the attached flesh.

Understanding Long-Bone Modification

Long-bone shaft fragments are a key to understanding the modification to these elements. In the Pod hradem assemblage, there are usually more shaft fragments for any given long limb element than there are whole shaft portions or complete bones—not an unusual phenomenon. This is evidence that at least some of the adult long bones have had their epiphyses removed, and that some have subsequently been broken apart by the carnivores. However, in contrast to the expectation of, for example, Klein (1975) and Marean and Spencer (1991) that shaft fragments would be better estimators of element abundance, there is no consistent similarity between the MNE estimated from shaft specimens, and that estimated from the most frequent surviving epiphysis (see Table 6.4). Thus, most of the fragments, of which there are in fact absolutely few, do not add significantly to the total MNE for any element.

Given that destruction by carnivores of large long bones proceeds from epiphyses inward, it is to be expected that many more shaft fragments than shaft cylinders would survive if utilization had not consistently deleted whole bones. Instead, in Stratum 17 complete shaft cylinders are represented more often than not, suggesting possibly light hyena, but more likely heavy wolf modification. An assemblage heavily utilized by hyenas would have deleted whole bones, most likely leaving only shaft shrapnel. In fact, the sample contains barely enough adult long-bone shaft fragments of each element to have made even one complete shaft for that element, an indication that these long-bone shafts were breached only rarely. Thus, I infer that either disinterested hyenas or very hungry wolves were involved.

We know from the percent-survival figures in Chapter 5 that all elements have survived in fewer numbers than would be expected based on MNIs. In light of this, using the relatively low numbers of shaft fragments as evidence of light modification by hyenas presents an ambiguous picture. What explains the absence of large numbers of whole elements, if they were not being heavily utilized? There are at least three possibilities that must be addressed. First, the modifying actors may have used some parts of the carcass heavily (the innominate and scapula, for example), while only lightly impacting others (such as adult long bones); second, that some whole elements were being transported out of the cave; third, that the missing elements are still buried in the unexcavated portion of the cave.

It would seem counterintuitive to imagine that the missing elements had been more heavily utilized or devoured in their entirety—surely there would have been more shaft fragments in such an assemblage. Moreover, degree of utilization results from, usually, systemic, ecologically driven constraints, such as density of predator population, and availability of carcasses. It is unlikely that certain parts of a given carcass would have been utilized to different degrees. It is much more likely that, at times, whole carcasses were treated differently than at others.

Given what is known of hyena and wolf behavior, transport out of the cave is a real possibility. However, the presence of smaller specimens provides evidence that argues against transport as the explanation for the missing specimens. Transport away from a carcass usually occurs as a reaction to intraspecific competition. In contrast to what one might expect if such competition had occurred, there is heavier utilization of subadult bones than the adult limb bones discussed above, including extremely high frequencies of gnaw marks and evidence of considerable time spent gnawing, resulting in polish and rounding on many specimens. This speaks of individuals spending time in the cave gnawing subadult bones, repeated over many episodes. An image of some animals with the leisure to gnaw some specimens rather intensely, after the fashion of a wolf homesite, does not accord well with the image of other animals being constrained at the same time to transport some of the larger carcass parts a distance from the death locale. In short, transport seems the weakest of the three arguments to explain the missing elements in this sample.

The mystery is that if hyenas were present often enough to have been responsible for destroying the numerous missing adult long bones, what would explain the existence of so many subadult specimens with only their epiphyses removed? And what explains the relatively high incidence of gnawing on all bones, and the ubiquity of gnaw marks on subadults?

Indeed, because adult elements were little utilized, and subadult specimens more heavily utilized, I am persuaded that wolves, less capable of modifying robust bones than hyenas, were responsible for the majority of modification to this assemblage. Heavy utilization by either hyenas or wolves might be expected to have deleted whole subadult bones. Yet, the predominant portion of subadult long bones is the shaft. It may be that the lack of desirable marrow in subadult bones would have resulted in only partial consumption of those bones at any degree of utilization by either hyenas or wolves. However, high numbers of gnaw marks on the subadult shafts strongly suggests that wolves and not hyenas were responsible. The reasoning behind this conclusion is as follows: a hyena would have needed little energy to remove the ends of such small bones. Thus, whether lightly utilizing the subadult carcass, or selectively removing the epiphyses because the diaphyses contained no desirable tissue, a hyena might not have gnawed the diaphyses to any great extent. Yet, in every case surviving subadult long bones are heavily gnawed, suggesting that considerable effort and time were required to remove the ends. Thus, they seem to have been treated in a manner analogous to the wild wolf "kennel" or homesite pattern observed by Haynes (1982)—extremely high rates of tooth marks, and break margins rounded by licking and contact with cave sediments.

In addition to the damage to subadult long bones, adult baculum and patella have been heavily gnawed in all cases. These are two elements that might not otherwise have interested hyenas, or (in the case of baculum) would have been devoured completely. Thus, it is reasonable to conclude that wolves, less capable of consuming bone than hyenas, were responsible for the majority of modification to this assemblage.

Finally, that the missing elements could be in the unexcavated portion of Stratum 17 is difficult to assess, but still a tantalizing possibility. The question of how to analyze the specimens that remain in such cases is a persistent problem in any archaeofaunal analysis based on an incomplete sample of a depositional unit, and one that is rarely discussed (but see Villa and Delpech 1993). However, given the slight damage observed above, and therefore the implausibility of transport, that the missing specimens might be on the other side of Pod hradem Stratum 17 seems, at least, possible. One line of evidence that may be used to assess this possibility is the survival of ribs and vertebrae. As previously mentioned, wolves and hyenas tend to leave these parts at carcass consumption sites. Certainly, one would not expect more than half of these elements to have been transported. Thus, with survival of vertebrae and ribs hovering around 50%, and with just about half of Stratum 17 having been left unexcavated, there seems good reason to suspect that, indeed, the ribs and vertebrae, and thus the remaining missing elements would be found in the unexcavated portion of the cave.

Detailed examination of the portional survival of individual elements has resulted in the production of a somewhat ambiguous dataset. All elements have undergone significant deletions of parts, most are gnawed to some degree, and some are virtually absent. There is evidence for wolf and hyena utilization of the bear carcasses, with the weight of evidence seemingly in favor of wolf modification. Although transport of carcass parts out of the cave seems unlikely, it is possible that the missing specimens could be in the unexcavated part of the cave.

Summary

In order to identify processes responsible for modifying bones in the past, it is necessary to be aware of present-day processes that can result in modification similar to that under study. By employing analogical reasoning, and recognizing relevant similarities between present and past phenomena, it is at least theoretically possible, and epistemologically defensible, to construct knowledge of the past in this way (Gifford-Gonzalez 1989, 1991; Wylie 1982, 1985, 1989a). There will always be the potential for more than one process to result in a given characteristic. This potential for the equifinality of different processes motivates archaeologists to continue expanding knowledge of bone-modifying processes under a variety of conditions, to tame, or at least to constrain the ambiguity that results, and thus to strengthen archaeological inference.

Evidence of carnivore modification in the Stratum 17 sample leads to the conclusion that both wolves and hyenas utilized the carcasses of cave bears in Pod hradem. Hyenas were undoubtedly responsible for the handful of shaft fragments remodelled by gastric juices (see Figure 6.5). Hyenas were also, most likely, responsible for the few adult mandibles that had their posterior inferior portions removed (Figure 6.8). There is, however, no evidence for hyenas heavily utilizing the bear carcasses, otherwise there might have been many more adult shaft fragments bearing evidence of cracking to remove the marrow. Instead, adult shaft fragments are few in number and mass, having few conchoidal flakes removed, and with shaft cylinders much more common. Thus, hyenas were probably infrequent visitors to the cave.

Evidence for wolf utilization, on the other hand, seems to be ubiquitous, with, at times, evidence for some very small individuals gnawing the smaller, subadult elements. The vast majority of tooth marks on cortical bone are less than 5 mm wide, and many that have scored the lamellae are on the order of 1 or 2 mm in width. Equally small and sharp are the numerous tooth punctures on the patella and baculum, two very dense elements. Hyenas may not have bothered with the patella, nor would they have gnawed the baculum without fragmenting it and, probably, devouring it, too.

In attempting to establish whether hyenas or wolves were the predominant modifiers of this assemblage, several factors were considered in concert. First among these was the unmistakable evidence, based on the size of tooth marks, that I have just alluded to. Second was the evidence for degree of utilization. Hyenas that had heavily utilized these bears would have left very little behind—some shaft fragments, but few epiphyses; some parts of the adult cranium. With the exception of the subadult long bones, which are commonly left as cylinders in spite of their small size, hyenas might have left much less of the subadult skeleton. In addition, the cranial bones of subadults would have been devoured entire. Instead, subadult cranial bones are the most commonly occurring. If hyenas were regularly impacting these bears' carcasses, it seems unlikely that they would have left so much of the subadults. Thus, it is defensible to say that the evidence points to wolves as the predominant modifiers of bone at Pod hradem Stratum 17. It must be acknowledged, however, that if information about transported parts and unexcavated remains were available, it could reinforce this conclusion, or it might force a different one. There is also the high probability that, due to the inherently variable nature of carnivore behavior with respect to changing ecological conditions, it might never be possible to distinguish unequivocally a dominant modifying actor at Pod hradem Cave.

Thus, I move to the next part of this study, the analysis of the spatial distribution of bone, with something less than a clear picture of what behavior is responsible for its characteristics. We know that bears died in the cave. We know that their carcasses were utilized by scavengers and their bones moved around at the whim of at least two species of bone-consuming carnivores. We know that afterwards other bears seeking refuge entered the cave to make nests, and further displaced or fragmented the bones of previously deceased and utilized conspecifics. Under certain conditions, those bears died, too, repeating the cycle.

With the knowledge of so many potential agents of disturbance having had a hand in the disposition of bones in Pod hradem Cave, it seems most unlikely that any coherent spatial patterning would be discovered. However, the next chapter reveals that some horizontal spatial clustering does occur, and that the patterning may, after all, reveal something about the habitual behavior of bone-consuming carnivores, as well as that of hibernating bears. Finding such patterns may, in the end, pose questions about the ability to infer pre-modern hominid cognition from the spatial distribution of bones in their caves.

CHAPTER 7

Spatial Analysis

The earlier chapters were concerned with narrowing the range of processes likely to have been responsible for forming the accumulation of cave bear and other bones in Stratum 17 at Pod hradem Cave. The evidence that emerged strongly suggests that during Stratum 17's deposition, Pod hradem Cave served as a scavenging theater, where wolves and hyenas utilized the carcasses of bears that had died during hibernation. In this second part, I explore the horizontal relationships of bone specimens in the cave, to determine whether the bears or postmortem disturbance by scavenging carnivores, or both, created any nonrandom horizontal distributions.

Theoretical and Methodological Background

The primary theoretical justification of this investigation is the need to further specify the range of processes forming spatial patterning in archaeological sites created by pre-modern hominids. As explained in Chapter 1, archaeologists sometimes use spatial patterning to make inferences about pre-modern hominid behavior. Models used in these interpretations are drawn from experience in the present, including ethnoarchaeological observations of modern human space use (e.g. activity areas, tool kits). As Chapter 1 argued, spatial patterning is of potential importance to the study of modern human cognitive evolution. Because of this, comparison with something other than modern humans is crucial to the development of an understanding of the determinants of spatial patterning in hominid sites, and the relationship between hominids, culture, and space. However, there has been little attention paid to discovering whether or not spatially patterned materials are truly unique to hominid sites. That is why I have chosen to explore the horizontal distribution of bones in a non-hominid locality.

In undertaking this exploration of the spatial distribution in a cave bear accumulation I had few well-developed expectations of the kind of spatial patterns that might emerge. Kurtén (1976) suggested a number of possibilities, based on his years of experience with cave assemblages, and on the writings of Koby (1941, 1953). For example, fragile elements coming to rest in out-of-the-way areas would likely be preserved, which might then be viewed as anomalous. More specimens might be expected to accumulate near the walls, where they would ultimately come to rest as the result of "housekeeping" on the part of the cave's inhabitants. Such was the case, most likely, in the Drachenloch Cave, where Bächler (1921) inferred that a cluster of cave bear crania in a natural niche of tumbled boulders represented purposeful behavior on the part of Neandertals, including the construction of an artificial container for the crania, and from which the "cult of the cave bear" was born (Kurtén 1976).

Kerbis-Peterhans (1990), in a study aimed at gaining a better understanding of the transport behavior of, among other animals, hyenas, observed clustering of bone specimens, long bones near to and oriented parallel to walls, bone-free pathways, and latrine areas in a hyena den in Israel. His observations, coming, as they did, early in my own research program, reassured me that my search for spatial patterning in a Pleistocene cave bear den might not be in vain. Kerbis-Peterhans's is the only archaeological account of such patterning of which I am aware. Lam's (1992) account of prey-choice variability among spotted hyenas mentions the tendency for cranial elements to accumulate inside the den he studied in Koobi Fora, Kenya. There were, he noted, many more limb bone elements in the accumulation outside the entrance. Thus, to date there has been sparse, though positive, evidence of the ability for animals to spatially pattern the bones in their accumulations.

The remainder of this chapter describes the results of my exploration of Pod hradem Stratum 17 for what it can reveal about the propensity for animals to structure the materials in their occupations. Eight facets of the sample will be examined, including patterning along taxonomic, size, and age dimensions.

Limitations of the Data

The excavators of Pod hradem Cave took pains to record the provenience of each and every bone specimen they unearthed. As a result, I have been able to perform qualitative and quantitative analyses of the spatial distribution of the cave bear and other fauna in one extensively excavated depositional unit. However, although the excavation of Pod hradem Cave involved the use of a one meter grid and 10 cm arbitrary levels, there are limitations to the quality of data retrieved that require elaboration.

As mentioned in Chapter 3, at times the excavators removed a 10 cm level over two or more contiguous 1 x 1 meter squares. So, for example, a sub-assemblage of specimens may be provenienced as "m^2 30-31," or "m^2 48-52." This makes it impossible to use unadulterated numbers in a spatial analysis based on grid counts—I required a count from each 1 x 1 m square. To render the observations useful for spatial analysis and to preserve the integrity of the data, I divided the total number of variates in a multi-unit block by the number of units contributing to the total. These averaged unit counts are what I employ in all spatial analyses. Both to adhere to minimal standards of data publication, and to enable future scholars to replicate my analyses, the original and the averaged data used in each spatial analysis are contained in Appendix 2.

I realize that, because some of the unit values are extrapolated from other numbers, any clustering that may have been present in the original deposits will be blurred by my procedure. It is, however, the only way to transform some of the observations made by the Moravské Museum into data useful for my purposes. In spite of this, I do not feel that my results will be invalidated by these averaging transformations. Since I seek patterns in the data, and the procedure I have adopted has the effect of evening out any uneven distributions, any clustering that is still perceptible must therefore be real. The presence of clustering after this averaging step implies that even stronger clustering may have been present originally, and which has simply been obscured by some irregular recording methods. Thus I do not think this procedure unfairly biases the data in my favor.

Some archaeologists may question the methodological wisdom of treating all the bones from a 1.0 meter thick depositional unit as belonging to one sample. However, I do not believe it is always possible, even with exacting excavation techniques, to isolate "floors" or "occupation levels" in so complex a depositional environment as a cave. There is simply too much latitude for postdepositional reworking of sediments in caves to expect that vertical relationships would be preserved at all times and in all places. Although it may not be possible in many cases to identify contemporary artifacts, this does not preclude the perception of patterning that transcends individual depositional events or occupations.

I expect that, in any depositional environment where successive occupations occur almost one on top of the other, time would tend to blur, or render apparently random, any patterns created by an actor whose behavior changes through time, as would be the case in many modern human populations. Thus, if patterning can be found in a deposit that built up over many years or centuries, it would make the argument of behavioral regularity that much stronger. I would argue that observations from Pod hradem can be used to study the habitual behavior of a species or several species of animal. If the behavior of a species is more or less constant through time, then gradual accumulation over a period of many years should not erase the physical traces of patterning. Trampling would only be expected to displace objects vertically, and thus should pose no threat to the horizontal integrity of patterning (cf. Gifford-Gonzalez et al. 1985). I am therefore confident that any patterning discerned in Stratum 17 at Pod hradem Cave must be the result of repetitive behavior on the part of the cave's occupants during the period of deposition.

Spatial Patterning in Pod Hradem Cave

This section explores some of the parameters of the horizontal distribution of bone specimens in the Pod hradem Cave Stratum 17 sample: the overall distribution of specimens, the distribution of specimens according to their physical dimensions (i.e. specimen length, volume, weight), distribution of cave bears according to age (i.e. adult, subadult, and fetal/neonate), and the distribution of the transported component of the assemblage.

Overall Distribution

In a cave bear hibernation locality there can be few expectations of the spatial distribution of bone specimens, because our knowledge of the behavior of these animals is limited to what can be inferred from the fossil record, what can be modelled from experience with modern bears, and by expectations of the physical processes that might be operating in a confined space, such as a cave. For example, if no peri-

or postmortem disturbance had occurred, we might expect the individual bones of whole skeletons to be found in close proximity to one another. However, we know that there was significant postmortem disturbance at Pod hradem, both by scavenging carnivores and, presumably, by subsequent ursid occupants preparing spaces to hibernate. If deposition in the cave was slow, there might have been occasion for considerable horizontal movement of large, obtrusive objects through trampling, and housekeeping, both on the part of bears and scavengers. Preparation of a comfortable hibernating place would have involved removal of large objects away from the chosen spot. Over time, these large objects would likely have ended up against a wall, away from which movement would have been unlikely. I thus expected to find more large specimens close to the wall of the cave. All else being equal, this should have produced a distribution having higher specimen counts in excavation units near the wall.

Given Kerbis-Peterhans's (1990) recovery of a cluster of gnawed bones in a modern hyena den, together with the presence of larger long bones against the cave's wall, I expected to find at least one nonrandomly occurring cluster that might have been due to the repetitive choice of a particular area of the cave, and, once again, larger objects near the wall. The corollary of this, areas with relatively few specimens, would also be expected, given the same behavior on the part of bears and scavengers. The statistical model against which this and subsequent distributions will be compared is one of random occurrence. If animal behavior in Pod hradem was capricious, the distribution of specimens should be random, especially considering the likelihood that Stratum 17 accumulated over a considerable time. Only repetitive behaviors, over the time Stratum 17 was building up, would or could produce patterns in the spatial distribution of specimens.

Figure 7.1 illustrates the distribution of the 2,402 specimens in Stratum 17's sample. High specimen counts occur in the two squares nearest the entrance, in a cluster near the wall at the back of the first excavation area, and in another cluster near the back of the second excavation area. There are strong density gradients in the distribution. Notice that the square in the bottom left contains 196 specimens, while nearby, there are units containing 10 or less. In addition there are a number of units that contained no specimens.

Upon inspection, there appear to be more high counts nearer the wall, with more moderate counts in the units that are more than a meter from the wall. The lowest counts occur more often at the edge of the excavation furthest from the wall. I first wanted to determine if this distribution could have occurred due to chance. Using Kintigh's (1985-1994) software, I calculated the exact binomial probability of

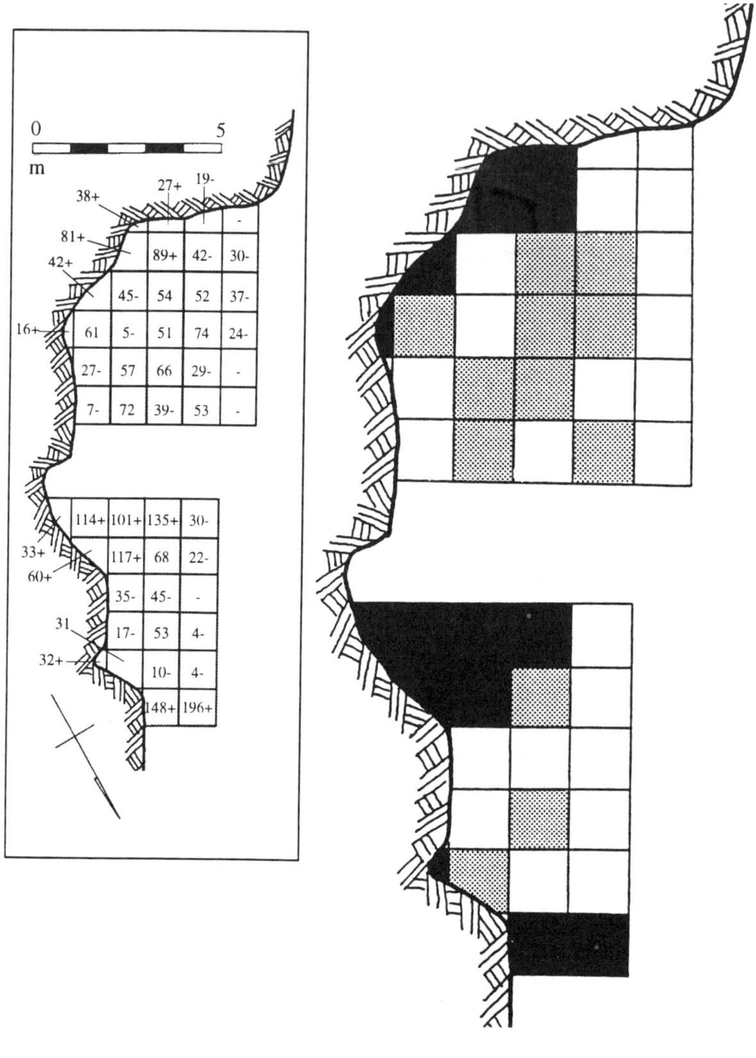

Figure 7.1. Pod hradem Cave Stratum 17: Horizontal distribution of all specimens. High counts with probability of occurrence ≤0.01 marked with "+"; low counts with probability of occurrence ≤0.01 marked with "-." On right, squares identified as having improbably high totals are filled; gray squares denote counts with a higher probability of occurrence (i.e. indistinguishable from random distribution); squares having improbably low totals are empty.

each count. Before doing so, however, I normed the counts for part squares to more accurately assess the relative density of specimens per unit. This was accomplished by multiplying the actual total by a factor based on the difference between the area covered by the excavated portion of the wall square and a 1 x 1 meter unit. The extrapolated totals and the exact probabilities for each excavation unit in this and each of the distributions subsequently examined in this analysis are given in Appendix 2. In Figure 7.1 those cells with improbably high counts (i.e. $p \leq 0.01$) are marked with a plus sign; those with improbably low counts (i.e. $p \leq 0.01$), with a minus sign. It seems as if more units with plus signs appear near the wall, and those having minus signs occur more often along the edge of the excavation, away from the wall.

To determine if such a distribution of abnormally high and low densities could have come about due to chance, I calculated the exact probability of finding so many high counts near the wall, once again using the binomial probability distribution. The likelihood of so many (i.e. 15) improbably high densities occurring near the wall, given the number (i.e. 19) of wall units, is $p=0.0042$. For improbably low densities, the probability of 10 occurring on the edge of the excavated area, given 11 units on that edge, away from the wall, is $p=0.0209$. Although this latter represents a slightly higher probability than that of so many high densities occurring near the wall, both the positions of the high counts and the placement of the low counts are extremely unlikely outcomes as compared with a random distribution. Thus, the subjective impression of clustering in the overall distribution of bone specimens at Pod hradem is statistically significant.

Given knowledge gained from the taphonomic study of this assemblage presented earlier, there is compelling evidence for the conclusion that animal behavior of some kind created nonrandom distributions in Stratum 17. However, the exact structuring processes have yet to be revealed. High specimen density in the entrance tunnel, for example, may be due to differential comminution from trampling, or may indeed represent an unusually high number of complete specimens. The likelihood of disturbance due to animals entering and leaving the cave would have been much greater in the narrow entrance tunnel, thus leading me to believe that comminution might be the causal factor. Other high concentrations, such as those near the walls further in, may have been protected from trampling, and may not be an artifact of fragmentation. Places where few bones occurred, the "negative" space nearer the back of the main hall, or along the edge furthest away from the wall, for example, cannot be attributed to trampling, and may therefore be due to some aspect of purposeful behavior, either on the part of the bears or the scavengers. Perhaps they avoided the most open spaces, preferring to remain nearer to the wall.

As noted above, Kerbis-Peterhans (1990) reports a cluster of hyena-gnawed bone in the cave at Arad, suggesting that carnivores prefer certain locations for certain activities. Thus, concentrations may be expected to result from similar behavior acted out by an unknown number of individuals over time. It also seems evident that at Pod hradem Cave several processes were at work patterning the overall distribution. There, carnivore scavenging and the spatial patterns it creates may be superimposed on patterns caused by preferred hibernation locations. Given the early stage of this analysis, it is not clear what the patterns in the overall distribution mean.

To begin to better understand the processes responsible for nonrandom distributions in Stratum 17, I looked at the physical parameters of the specimens. If animals were creating the patterns by displacing obtrusive objects, analysis of specimen length, volume and weight might prove illuminating.

Specimen Length

Partly because I was searching for any patterning that might exist, and because of the anecdotal evidence provided by Kurtén (1976), Koby (1941, 1953) and the archaeologically recovered evidence of Kerbis-Peterhans (1990), I looked at the physical dimensions of the Pod hradem sample. One of my expectations was that differently sized pieces would have been treated differently by hibernating bears—they might not have wanted to spend six months lying on a complete cave bear femur, for example. Observation of modern bears reveals that they prepare hibernation places by introducing vegetation, probably to act as a moisture barrier and as a cushion (Rogers 1981). Thus, it seems plausible that cave bears pushed aside large bone pieces to make a comfortable hibernating place. In a confined space, such as a cave, repeated displacement of large objects might result in their coming to rest near walls. Once there, unless purposely removed, they would be less likely to find their way out into the open again. This could result in a disproportionate number of obtrusive specimens near, or against the wall.

To obtain an idea of whether or not specimen length is related to the spatial distribution, I looked at the average length of specimens per unit. My analysis revealed that the frequency distribution of specimen length is highly skewed to the left, with the arithmetic mean at 70 mm, the most common value at 40 mm, and half the collection at 55 mm or less in length. Many specimens, however, are significantly larger than the modal size. Notice that the two squares nearest the entrance have very low average lengths. This is one indication that, indeed, trampling has resulted in the comminution of the bones in those squares, leading to their having abnormally high specimen counts (see Figure 7.1).

To assess the likelihood that the distribution of specimen average lengths shown in Figure 7.2 could have been due to chance, unit averages were compared to a binomial probability distribution. Improbably high (i.e. $p \leq 0.01$) averages are marked with a plus sign; improbably low averages are marked with a minus sign. Immediately one can see that the majority of high averages occurs next to the wall. And, once again, most of the improbably low values occur furthest away from the wall. To decide if this distribution of high and low average lengths could have been due to chance, I calculated the exact probability, using, once again, a binomial distribution.

The probability of getting 6 of the 8 improbably high totals against the wall is $p=0.01$. This is an extremely unlikely occurrence, and strongly supports the notion that longer items occur near the wall. In a similar fashion, I calculated the probability of getting so many low counts furthest away from the wall. The probability of 6 of 7 occurring furthest away from the wall is $p=0.001$. Once again, this is an extremely unlikely situation, and strongly suggests that this distribution is nonrandom.

Some process or combination of processes resulted in above-average-length specimens occurring at the wall most of the time. The observations of Kerbis-Peterhans (1990), Kurtén (1976), and Koby (1941, 1953) suggest that this kind of size sorting is the result of "housekeeping" on the part of occupants. Given the understanding of present-day denning bears mentioned in Chapter 2, such "housekeeping" may not be out of the question. The preparation of a "comfortable" sleeping place may be relatively more important if one is sleeping for six months at a time.

Whether it was the wolves, the hyenas, or the hibernating bears alone, or simply the combination of their behaviors, there is a strong tendency for large bones and fragments to occur near the wall in Pod hradem Cave Stratum 17, implying that animal behavior was responsible for the patterning. Likewise, "empty" spaces, which occur furthest away from the wall, suggest that a tendency to avoid those areas may have been responsible.

If, indeed, animals differentially move extremely long objects out of their activity areas, items that take up a large area might also be expected to undergo the same treatment. A bone that is not very long

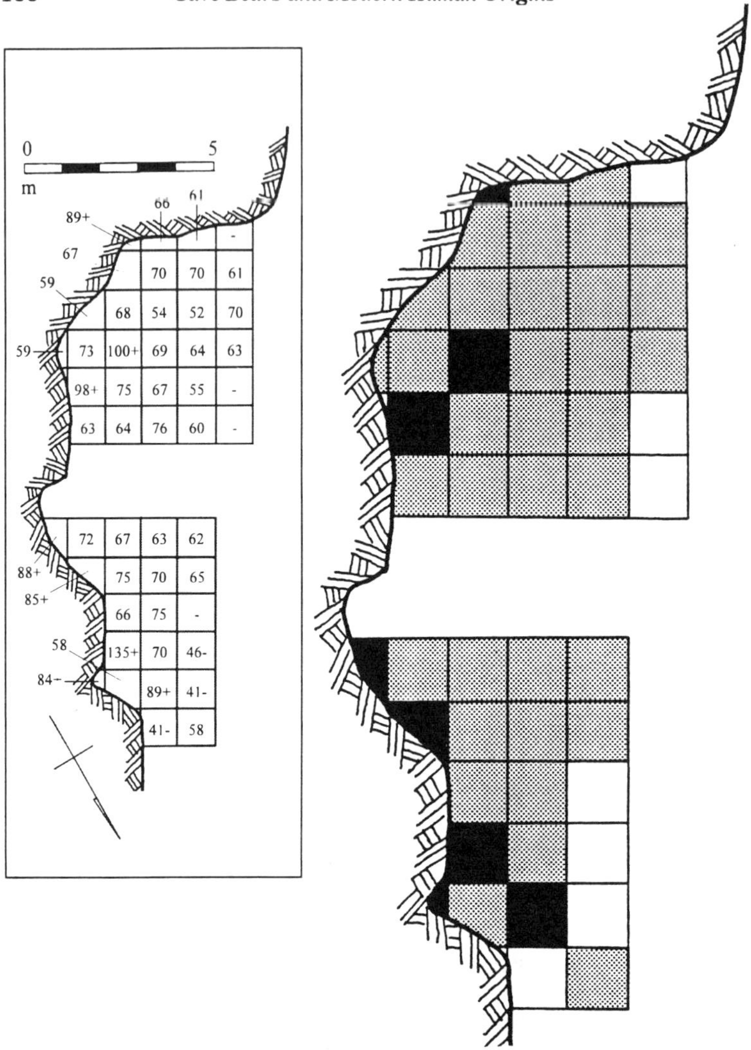

Figure 7.2. Pod hradem Cave Stratum 17: Horizontal distribution of average specimen length (mm). High average lengths with probability of occurrence ≤0.01 marked with "+"; low average lengths marked with "-." On right, squares identified as containing improbably high average specimen lengths are filled; gray squares denote average lengths with a higher probability of occurrence (i.e. cannot be distinguished from random distribution); squares having improbably low average lengths are empty.

may still be obtrusive—for example, an adult sacrum or large vertebra. To see if something we might call "overall obtrusiveness" contributed to the patterning at Pod hradem, I obtained the product of length and width of each specimen, and then plotted the average area of specimens in each excavation unit. Figure 7.3 illustrates the results. Once again, I calculated the exact probabilities for each unit's value, and once again improbably high counts are denoted by plus signs; improbably low by minus signs. Although the patterning is less robust than in the example of average length, it is still possible to observe the same process at work. The three highest per unit averages (i.e. 115, 80, and 61 cm^2) occur against the wall. Although the distribution is not statistically significant, as in previous cases, at least for a part of the excavated portion of the cave the most obtrusive objects do end up against the wall in most cases. Also, as was the case with average length, few obtrusive objects occur in the open area of the cave.

Thus, obtrusive objects form a slightly different distribution than that of long objects, although there is broad similarity between the two. This makes sense, since one of the components of the area calculation was length. Nevertheless, specimen area provides evidence of another nonrandom pattern in the accumulation at Pod hradem.

Specimen Weight

As one would expect in an assemblage where the length of specimen in some ways determines its disposition, and where the most obtrusive objects, in general, are patterning this assemblage, one would think that the weight of an object might also be used to track the processes forming the patterns—because the more dense the object, and the larger the object, the more the object weighs. I thus looked at weight as a proxy for "obtrusiveness." The results are presented in Figure 7.4. It is obvious that some of the cells with the highest counts occur nearest the walls.

To discover if this distribution of unusually heavy objects could have come about by chance, I calculated the exact probabilities, using the binomial distribution, of the average weight per unit. Improbably high counts, once again, are denoted with plus signs; low counts with minus signs. There seems to be a tendency for very low averages to appear furthest away from the wall, as one might expect given the low average lengths in that area. However, this analysis produces the greatest number of improbably low counts of any of the three dealing with physical properties. The probability of getting so many improbably high average weights in cells near the wall is $p=0.036045$. Although this is not conclusively nonrandom, it is nevertheless highly

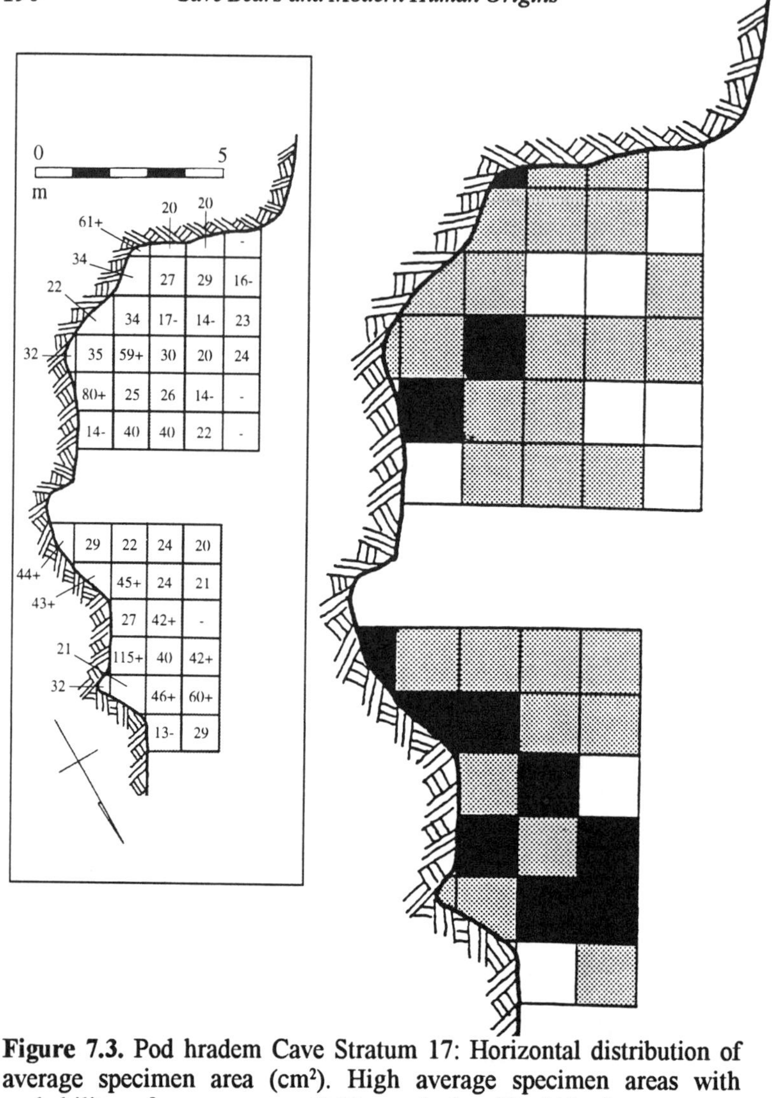

Figure 7.3. Pod hradem Cave Stratum 17: Horizontal distribution of average specimen area (cm^2). High average specimen areas with probability of occurrence ≤0.01 marked with "+"; low average specimen areas with probability of occurrence ≤0.01 marked with "-." On right, squares with improbably high average specimen area are filled; gray squares denote average specimen area with a higher probability of occurrence (i.e. indistinguishable from random distribution); squares with improbably low average specimen areas are empty.

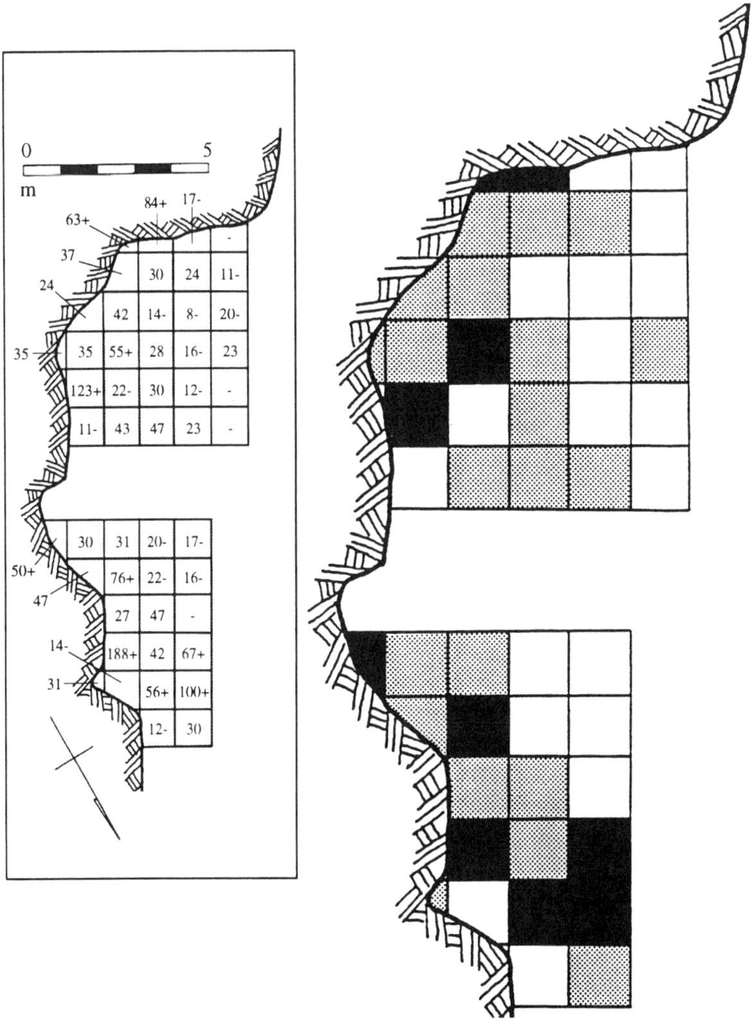

Figure 7.4. Pod hradem Cave Stratum 17: Horizontal distribution of average specimen weight (gm). High average specimen weights with probability of occurrence ≤0.01 marked with "+"; low average specimen weights with probability of occurrence ≤0.01 marked with " - ." On right, squares with improbably high average specimen weights are filled; gray squares denote average specimen weights with a higher probability of occurrence (i.e. indistinguishable from random distribution); squares with improbably low average specimen weights are empty.

unlikely. The probability that so many low averages would occur away from the wall is p=0.0000. These results are good evidence that regular, probably intentional behavior on the part of animals created unmistakable spatial patterns.

On the model of size sorting proposed by Binford (1978b), among others, many spatial analysts monitor the size of objects for what can be inferred about the activities of the humans that created an archaeological site. It is clear from this analysis of the spatial distribution of specimen size at Pod hradem that animals, too, produce size-sorted accumulations, for much the same reasons, presumably, as humans do—to get rid of obtrusive objects. Humans are cultural beings, but their cultural abilities need have no bearing on, at least, the size sorting component of spatial patterning. It cannot be claimed that the animals that created the patterning in Pod hradem were conscious of what they were doing when they moved the specimens around, in the way that modern humans are capable of being conscious of what they do. Nevertheless, in bear caves, or hominid occupations, whether preparing a comfortable place to hibernate, or just going to sleep for the night, the most obtrusive objects would likely be differentially relocated away from the chosen spot. It might be expected, moreover, that the best spots for reclining might stay the same through time. Through time, then, this might lead to obtrusive objects coming to rest out of the way of traffic and otherwise heavily used space. It should come as no surprise, under those circumstances, to find the larger pieces at the edges of the habitable zone, in this case the walls. There it is unlikely they would be moved, or broken up, further. Presently I can think of no other explanation for this distribution of large and heavy specimens (which includes the three relatively intact adult crania) in squares near the wall.

Distribution of Cave Bear

If bear behavior had any part in creating the distribution of bones in Stratum 17, the pattern should be discernible when only cave bear bones are analyzed. However, since the vast majority of specimens in the assemblage are ursid, there seemed no point in monitoring the bears altogether. Instead, and for reasons I will discuss below, I grouped the specimens according to the three age classes: adult, subadult, and fetal/neonate.

Adults. Perhaps as much as 80% of the total number of specimens in the assemblage are adult cave bear. Thus, the adult distribution should be expected to map on fairly closely to the overall distribution of specimens that was shown in Figure 7.1. Indeed, as Figure 7.5

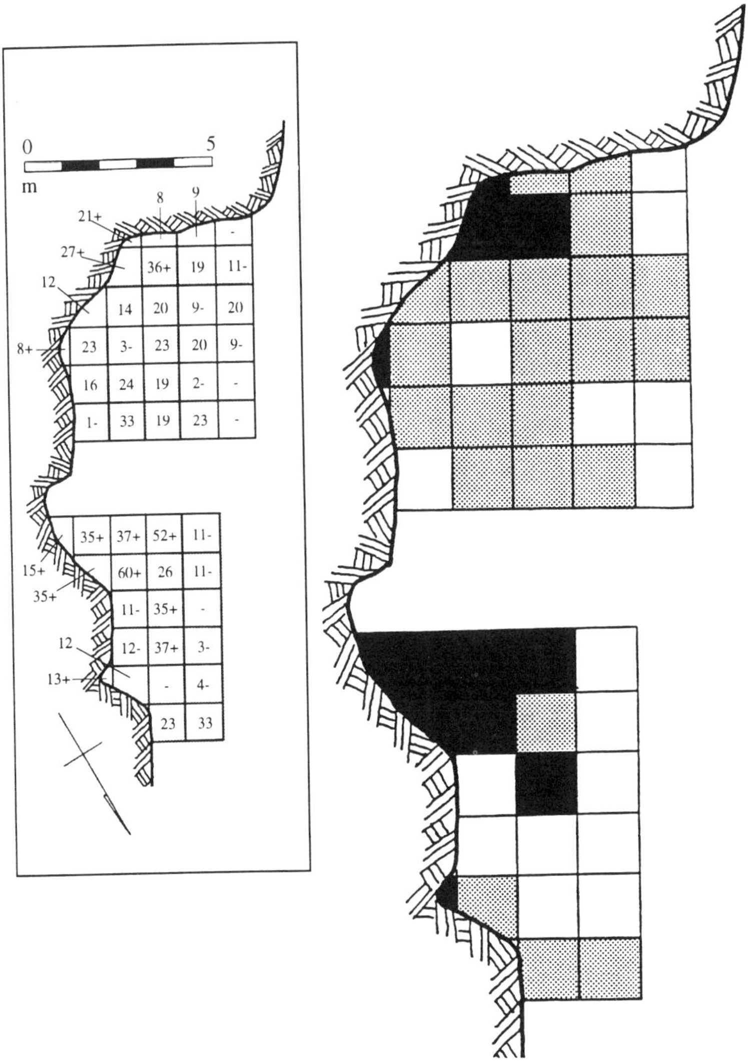

Figure 7.5. Pod hradem Cave Stratum 17: Horizontal distribution of adult *Ursus spelaeus*. High counts with probability of occurrence ≤0.01 marked with "+"; low counts with probability of occurrence ≤0.01 marked with "-." On right, squares with improbably high counts are filled; gray squares denote counts with a higher probability of occurrence (i.e. indistinguishable from random distribution); squares with improbably low counts are empty.

illustrates, adult specimens are ubiquitous. Comparison with a binomial probability distribution yields a pattern of high (plus signs) and low (minus signs) counts having low probabilities (i.e. $p \leq 0.01$) very similar to that in Figure 7.1. One main cluster of high densities occurs near the excavation break, and more low densities occur in the open area.

Subadults. When subadults are monitored, however, a slightly different distribution of improbably high units emerges than was the case for adult bears (Figure 7.5). Figure 7.6 illustrates anomalously high counts in one area where only moderate counts are to be found in the overall bone distribution shown in Figure 7.1. Yet, elsewhere in the cave where overall counts are high, subadult counts are only what might be expected if the distribution were random. This suggests, at least, that subadults were treated differently than adults, which might be expected, given the generally well-gnawed state of most subadults. If the scavengers had a preferred location, and if subadults were being preferentially gnawed, it would seem most likely that the subadult specimens would occur most often there. Thus, we can speculate that the area near the back of the first excavation block may have been the "epicenter" of scavenger activity. Indeed, the overall distribution of bones supports this inference (Figure 7.1).

Fetal/neonates. Finally, I examined the distribution of fetal/neonate specimens. Bones of fetal and newborn cave bears are on the whole very small (i.e. less than 50 mm on average, with long bones ranging in length from 10 to 15 mm). When they were not consumed entirely, either during utilization of the pregnant female, or shortly after birth, their bones might have been overlooked by scavenging carnivores. As such, trampling may have moved them only a little from the location of their death. Figure 7.7 presents the distribution of these bones. They, too, appear to be nonrandomly distributed. I calculated the exact probabilities of occurrence for each cell in the sample. Those excavation units with highly improbable occurrences (i.e. $p \leq 0.01$) are marked with plus signs if high and minus signs, if low. There appear to be relatively few fetal/neonate specimens near the wall, in contrast to the overall number of specimens, which tend to cluster near the wall. There also appear to be few of these small specimens in the open area immediately inside from the high concentration of specimens near the entrance, and again, in the central portion. High densities occur in the entrance tunnel, in the concentration of bones about midway along the left, and in a cluster near the back of the cave. This latter is almost entirely anomalous—there is no similar concentration in the overall distribution (see Figure 7.1). If, indeed, fetal/neonate bones were left

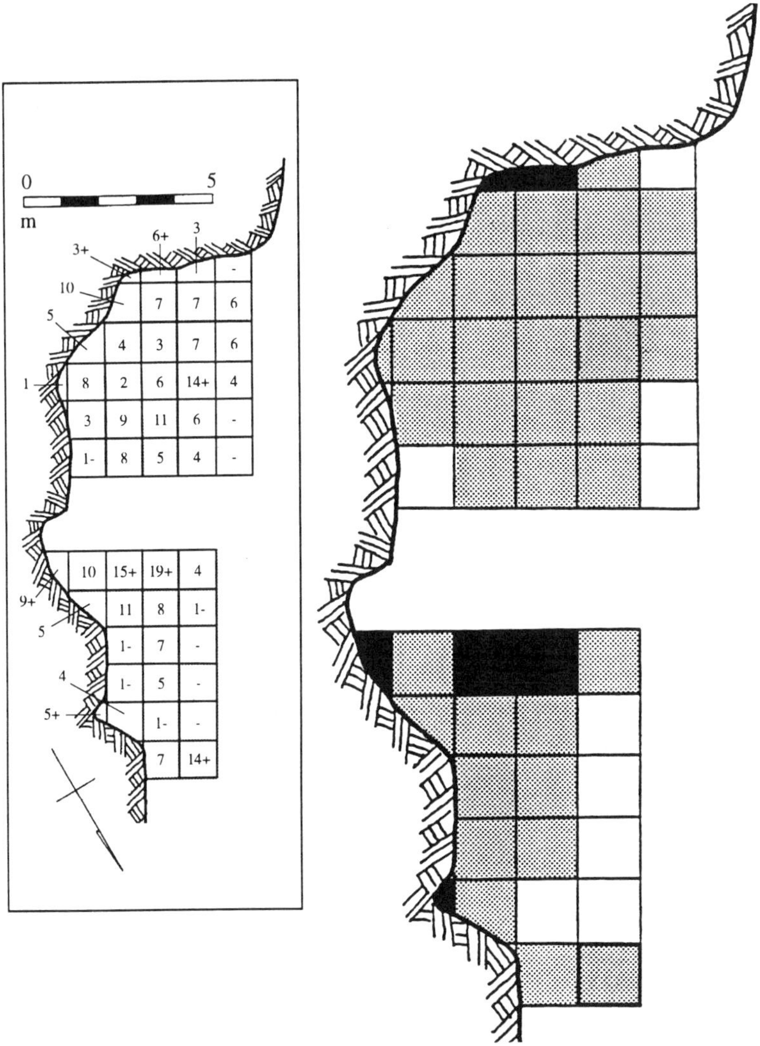

Figure 7.6. Pod hradem Cave Stratum 17: Horizontal distribution of subadult *Ursus spelaeus*. High counts with probability of occurrence ≤0.01 marked with "+"; low counts with probability of occurrence ≤0.01 marked with "-." On right, squares with improbably high counts are filled; gray squares denote counts with a higher probability of occurrence (i.e. indistinguishable from random distribution); squares with improbably low counts are empty.

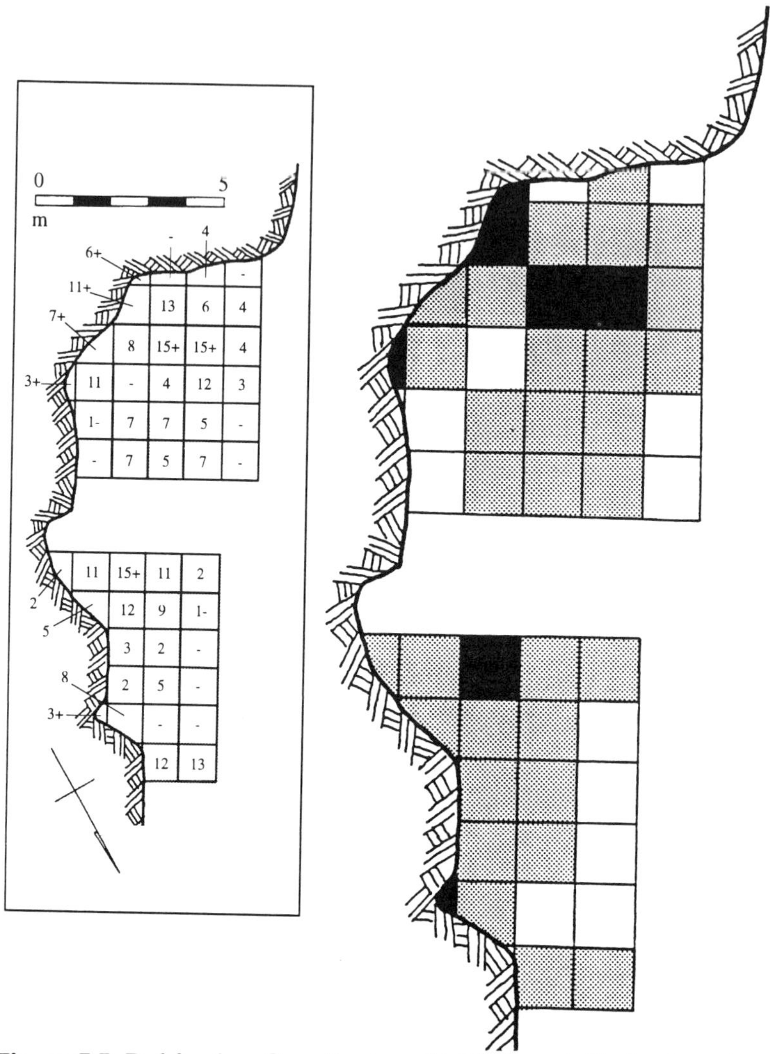

Figure 7.7. Pod hradem Cave Ştratum 17: Horizontal distribution of fetal/neonate *Ursus spelaeus*. High counts with probability of occurrence ≤0.01 marked with "+"; low counts with probability of occurrence ≤0.01 marked with "-." On right, squares identified as having improbably high counts are filled; gray squares denote counts with a higher probability of occurrence (i.e. cannot be distinguished from random distribution); squares having improbably low counts are empty.

alone, for the most part, these concentrations could represent natal areas—locations preferred by denning pregnant bears. Interestingly, the densest concentration is furthest away from the cave entrance. Perhaps this was an area perennially used by denning females.

Compared with the overall distribution of cave bear specimens in the cave (Figure 7.1), the anomalous distribution of fetal/neonates begs explanation. If carnivore behavior is responsible for the overall distribution, the explanation just proposed for the location of fetal/neonate bones makes good sense. This is where they were born and, presumably, died, to be disturbed no further until they were excavated by the Moravské Museum.

Distribution of Carnivore-Transported Specimens

Given the importance of determining whether or not the carnivores were responsible for most of the patterning in Pod hradem, I thought that it might be possible to monitor the effects of carnivores apart from the behavior of bears, by looking at where the transported (i.e. non-ursid) specimens ended up. I reasoned that looking at the non-bear component should be instructive, whether scavenging carnivores were primarily responsible for the bone distribution, or had simply overprinted their patterning on that of the bears. When the bones of carnivores other than bear are monitored, there is a hint that the distribution is patterned.

Figure 7.8 gives the distribution of bones that must have come into the cave through the agency of non-bear carnivores. Although there are few specimens securely identifiable relative to the total, there appears to be a tendency for them to cluster near the wall, as with the overall distribution. However, when the distribution is compared to a theoretical binomial distribution, the results are insignificant. This means that the arrangement of non-bear specimens cannot be statistically differentiated from the overall distribution. This can be used to suggest tentatively that the distribution of non-ursid specimens may have been the result of the same process that patterned the overall distribution. If the shape of the non-bear component is the same as that of the overall, and was itself the result of carnivore behavior, perhaps this represents proxy evidence for the source of most of the patterning, i.e. the scavengers. Thus, the distribution of non-bear specimens may indeed be an accurate picture of the areas preferred by the scavenging wolves and hyenas.

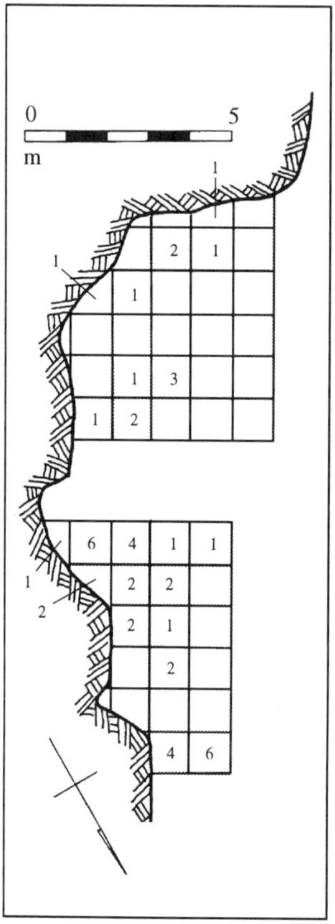

Figure 7.8. Pod hradem Cave Stratum 17: Horizontal distribution of non-ursid specimens. This distribution is not statistically distinguishable from the overall specimen pattern (see, e.g., Figure 7.1), nor do any of the counts vary from what might be expected due to chance.

Summary

An examination of the spatial parameters of Pod hradem's bone accumulation has revealed a number of nonrandomly patterned distributions. More bones overall, more large bones, and more heavy bones tend to occur near the wall. I have suggested that this is related to the repeated, perhaps random movement of obtrusive pieces, resulting in their coming to rest against the wall, where they would be less likely to be moved any further. Bears preparing sleeping places, perhaps, and carnivores removing carcass parts and subsequently losing interest, or both, could explain the pattern of distribution along these lines.

Although inconclusive, it has also been demonstrated that when carnivores introduced the parts of other animals to the cave, they tended to do so in a circumscribed area, suggesting that their behavior in relation to space is repetitive. Heavily gnawed bones, too, tend to occur in high numbers where high numbers of non-bear bones occur, suggesting a relationship between the carnivore-introduced bone and the heavily chewed specimens.

Thus, the bone accumulation at Pod hradem was undoubtedly spatially structured by the repetitive behaviors of the carnivores that scavenged bear carcasses, by cave bears preparing hibernation places, and by the preferential use of parts of the cave by denning female bears through time.

CHAPTER 8

Implications for the Archaeology of Modern Human Origins

> ... we stand today at the threshold of the last step in the study of spatial patterning—we are beginning to know what we do not know about the meaning of the distribution of prehistoric materials at stone age sites. In effect we are at the end of the beginning of spatial analysis studies. We know that patterns of distribution contain important information. We have the methods necessary to define such patterns. We are now beginning to concentrate on what those patterns mean.
> (Kroll and Price 1991:304-305)

Research into modern human origins is complicated by a lack of consensus on what it means to be human, what language is, and on what culture is. This has resulted in a stalemate, of sorts, between scholars who hold to one of two different models for the emergence of modern humans—one proposing multiregional origins for modern humans; the other, advocating widespread replacement of pre-modern hominids by behaviorally modern (and skeletally modern) humans. Chapter 1 argued that this may be the result of two conflicting notions about the cognitive complexity of pre-modern hominids in relation to that of modern humans. It was suggested that archaeological inferences of pre-modern hominid cognitive abilities may be compromised by unwarranted use of formal analogies having their basis in modern human behavior. The makers of Mousterian stone tools, for example, especially of handaxes, and the practitioners of the Levallois technique of manufacturing stone tools, have traditionally been viewed as possessing modern human technological, and thus cultural, complexity. However, ascriptions of modern human cognition to those same stone-tool makers have recently been called into question. Such ascriptions can be traced to formal analogies, drawn from modern human use of stone, and applied without adequate warrant to the pre-modern hominids whose behavior such stone-working represents (cf. Dibble 1987a, 1987b; Davidson and Noble 1993). As Chapter 1 concluded, in light of such examples, it is apparent that the intellectual stalemate on the question of modern human origins is at least in part the result of the unwarranted application of formal analogies, marshalled in the interpretation of pre-modern hominid behavior. If, indeed, assessments of pre-modern hominid cognition are open to question, then there is a need to further develop the means by which inferences of pre-modern hominid cognition are made.

This study has contributed to the construction of knowledge about modern human origins, through further specifying the range of processes that can create spatial patterns in cave bone accumulations. Until now there has been little rigorous effort to explore the ability of non-hominids to structure their sites spatially. My goal has been to discover whether or not non-hominids can create spatial patterns in cave bone accumulations—either through repeated movements, or through partitioning of space for different activities—to tease out the implications of finding spatial patterning in the sites of pre-modern hominids.

Previous chapters have examined the processes contributing to Pod hradem Cave's spatial distribution of bones. They are summarized below. Following that I offer some conclusions about what this research means to the archaeology of modern human origins.

Summary

The Pod hradem Cave bear fauna is considered to have been the result of cave bears dying during hibernation (Moravské Museum 1965). There is good reason to accept this conclusion, given the absolutely high numbers of cave bear remains, and the relatively high numbers of juvenile ursid remains. It was necessary, however, to investigate the possibility that the assemblage was the result of a combination of processes including transport of cave bear carcass parts, because 1) the skeletal material was not articulated, 2) the remains of non-ursid carnivores and other species occur, and 3) the majority of the specimens had been subjected to considerable carnivore modification. The goal of the taphonomic analysis was to elucidate the processes that created the bone accumulation, so that it could be said which of the processes contributed to any spatial distributions emerging from this study.

Taphonomy

To identify processes responsible for past bone accumulations, it is useful to understand present-day processes that can result in patterns caused by the same determining structures as those under study. By employing analogical reasoning, and recognizing relevant similarities between present and past phenomena, it is at least theoretically possible, and epistemologically defensible, to construct knowledge of past behavior in this way (Gifford-Gonzalez 1989, 1991; Wylie 1982, 1985, 1989a). However, there will always be the potential for more than one process to result in a given assemblage characteristic. This potential for the equifinality of different processes motivates archaeologists to continue expanding knowledge of site formation processes under a variety of conditions, to tame, or at least to constrain the ambiguity that results, and thus to strengthen archaeological inference.

Assemblage Composition and Age at Death. Knowledge of the species represented in Pod hradem enables one to begin to assess the processes that formed the site. We know, for example, that cervid and bovid remains would most likely have been transported into the cave by meat-eating carnivores, and not by the cave bears. On the other hand, wolves, hyenas, and bears are known to use caves as natal dens, refuge from competitors, and places to hibernate. Thus, their bones could be expected to occur as the result of deaths accompanying their presence in and uses of the cave.

Securely identified cave bear specimens comprise slightly over 60% of the assemblage, but that number is probably closer to 80%, if one takes into account those specimens identified with the most conservative criteria, and which are in all probability those of *Ursus spelaeus*. The majority of the securely identified non-bear taxa in Pod hradem are common prey of wolves and hyenas, whose bones are also found in small numbers in the cave. Thus, the possibility that Pod hradem functioned as a den for wolves and hyenas was entertained.

The results of analyses of species composition and age at death suggest that Pod hradem was, however, most likely a hibernation den.

Dental attrition data imply winter mortality peaks. Greater than twice as many cave bear juveniles are represented in the assemblage as adults. In addition, the absence of prime age cave bear adults creates a U-shaped mortality profile, which could have resulted either from winter (hibernation) mortality peaks in bear populations, or from the accumulation of carcasses hunted by cursorial predators, or a combination of both. The cave, however, would have presented wolves and hyenas, both cursorial predators, with a classic "ambush" opportunity, which should have yielded a mortality profile more like a living age structure. Thus, it can be fairly confidently stated that the Pod hradem Stratum 17 cave bears were not dying as the result of predation, but rather were scavenged. Finally, low numbers of non-bear teeth, only one tooth of a juvenile non-bear carnivore, and overall low numbers of non-bear remains in Stratum 17 do not accord with the expectations of carnivore natal dens. Taken together, these data suggest that during the accumulation of Stratum 17 Pod hradem Cave was probably not the natal den of non-bear carnivores, nor a place where carnivores preyed upon hibernating bears. However, after analysis of taxonomic representation and age at death, it was still not clear if considerable use of the cave as a non-ursid refuge den could be ruled out, or if in fact a significant component of the cave bear remains had been introduced by the other carnivores.

Anatomical Composition. Element frequencies provide another line of evidence in the search for the causal agent responsible for the characteristics of this bone assemblage. Element frequencies found at Pod hradem generally look much more like the accumulated skeletal remains of open-air wolf and hyena carcass-processing sites than those of their den sites. While on the surface paradoxical, the resemblance of a cave assemblage to that of open-air kill sites actually makes good sense. If the bears were dying during hibernation in Pod hradem, carnivores would have come upon the carcasses of recently dead bears and have utilized them according to the same habits typical of their handling of whole carcasses elsewhere in the landscape. If that were the case, high numbers of vertebrae and ribs should characterize such assemblages. By contrast, the contents of a carnivore den are normally comprised of those elements transported away from the death and consumption site, creating a mirror image of the death site in terms of anatomical composition. Thus, low numbers of vertebrae and ribs could be expected in hyena and wolf den sites. In fact, in Pod hradem's Stratum 17, high numbers of rib and vertebral specimens occur. Moreover, high and significant rank-order correlations between the cave bear element survival and that of wolf (and to some extent hyena) open-air carcass-consumption sites provides powerful evidence that the cave bear remains in Pod hradem Cave were not transported there by carnivores. They were, most likely, the result of hibernation mortality.

Modification. Despite the close concordance between adult cave bear anatomical composition and that of open-air carnivore carcass consumption, it was still necessary to observe the patterns of modification to better understand whether wolf or hyena were primarily responsible for the disturbance to Pod hradem's cave bear bone accumulation.

Evidence of carnivore modification in the Stratum 17 sample leads to the conclusion that both wolves and hyenas utilized the carcasses of cave bears in Pod hradem. Hyenas were undoubtedly responsible for the handful of shaft fragments remodelled by gastric juices and, perhaps, for the few adult cave bear mandibles that had their posterior inferior portions removed. However, hyenas do not appear to have lingered in the cave, or to have heavily utilized the bear carcasses, otherwise there might have been many more shaft fragments bearing evidence of cracking to remove the marrow. Instead, adult shaft fragments are few in number and mass, with complete elements and shaft cylinders much more common. Thus, hyenas appear to have made an insignificant contribution to the patterning seen in the sample.

Instead, evidence for wolf utilization seems to be ubiquitous, with, at times, evidence for some very small individuals gnawing the smaller, subadult elements. The vast majority of tooth marks on cortical bone are less than 5 mm wide, and many that have scored the lamellae are on the order of 1 or 2 mm in width. Equally small and sharp are the numerous tooth punctures on the patella and baculum, two very dense elements in association with, respectively, the tendons of a large muscle group, and of the easily accessible flesh in the anal/genital region. Hyenas may not have bothered with the patella, *per se*, nor would they have gnawed the baculum without fragmenting it and, probably, devouring it along with the flesh containing it.

In attempting to establish whether hyenas or wolves were the predominant modifiers of this assemblage, several factors were taken into account. First was the size of tooth marks. Second was the evidence for degree of utilization. Hyenas heavily utilizing these bears would have left some shaft fragments, but few epiphyses, and some parts of the adult cranium. They might also have left much less of the subadult skeleton. The cranial bones of subadults would have been devoured entirely. Instead, they are the most commonly occurring elements in the assemblage. Lastly, high numbers of tooth marks on subadult long-bone shafts effectively rules out hyenas as the predominant modifying actor, because they would not have found it necessary to "worry" such small bones—one or two bites would have collapsed a subadult bear long bone. Thus, the bulk of evidence points to wolves as the predominant modifiers of bone in Pod hradem Cave's Stratum 17.

Spatial Analysis

An examination of the spatial parameters of the bone accumulation at Pod hradem revealed that more bones overall, more large bones, and more heavy bones tend to occur near the wall. Bears preparing sleeping places, perhaps, and carnivores moving carcass parts and subsequently losing interest, could explain the pattern of distribution along these lines. I have suggested that this related to the movement of obtrusive pieces, resulting in their coming to rest against the wall.

It has also been demonstrated that when carnivores introduced the parts of other animals to the cave, they tended to do so in circumscribed areas, suggesting that their behavior in relation to space may be governed by habitual action. Heavily gnawed bones tend to occur in high numbers where the non-bear bones occur, suggesting a relationship between the carnivore-introduced bone and the heavily chewed specimens. Finally, the location of fetal/neonate cave bear specimens suggests that areas of the cave were preferred by pregnant females. Thus, it can be inferred that the spatial distribution of Pod hradem's bone accumulation was structured by the repetitive behaviors of carnivores that scavenged bear carcasses, and, probably, by the preferential use of parts of the cave by bears through time.

Throughout the previous seven chapters, I have maintained a level of optimism about the potential for spatial analysis to be useful in the search for understanding about the emergence of modern humans. I fervently hoped that this work would not be viewed as yet another "cautionary" tale from the taphonomic perspective. Rather, I hope that spatial analysis may possess a previously unrecognized potential for understanding pre-modern hominid cognitive abilities, as explained in the next section.

Implications for the Archaeology of Modern Human Origins

One of the obstacles in the way of research into modern human origins is that there is no consensus on what it means to be human, much less "behaviorally modern." Furthermore, there is no consensus among archaeologists and paleoanthropologists as to what culture is, or on how it is to be included in the scholarly pursuit of knowledge about the archaeological past. Much of the debate over when modern humans emerged can be traced to this lack of agreement on what culture is, what language is, and, in general, on what kinds of behavior are good evidence of modern human abilities.

This study has demonstrated that non-hominid behaviors produced spatial patterning in a bone accumulation. Together with the anecdotal evidence of Kurtén (1976) and Koby (1941, 1953), and the recent descriptions by Kerbis-Peterhans (1990) and Lam (1992), it can be suggested that such patterning resulting from animal behavior is not a rarity. The presence of spatially patterned traces in a non-hominid bone accumulation demonstrates that neither culture nor hominids are uniquely able to produce spatially patterned material residues.

Knowing that neither the hominid brain, nor human culture, are prerequisites for an organism's structured use of space should encourage paleoanthropologists to develop a more nuanced approach to spatial patterning in hominid sites, and to avoid the misplaced use of formal analogies based on modern human behavior in their interpretation. As I argued in Chapter 1, spatial patterning is often interpreted explicitly or implicitly by reference to modern human use of space and the patterning that results. Thus, finding spatial patterns in animal accumulations should tell us that by focussing on behavior without considering the motivation for it, we may be hampering our efforts to understand the transition from pre-modern hominid to human. Although similar *behaviors*—such as the desire to remove obtrusive objects—may underlie similar nonrandom spatial patterns in modern human, pre-modern hominid, and cave bear sites, we can be certain that those similar behaviors are not the result of similar cognitive processes on the part of modern humans and bears. Thus, this calls into question the nature of pre-modern hominid cognition as revealed in spatial patterning.

Having found spatial patterning in a non-hominid context, it may be useful to re-examine the methodological basis for spatial analysis of pre-modern hominid sites, and to develop alternative ways of interpreting the causes of spatial patterning in those sites. Indeed, for spatial analysis to advance understanding of modern human origins—i.e. to track the evolution of modern human cognitive abilities in hominids—those abilities must be inferred from the spatial patterns found in hominid sites. Simply inferring behavior will not be sufficient. Thus, to impute the cause of spatial patterning in any hominid site to modern human cognitive abilities, it will first be necessary to find the processual linkages between modern human cognition and the behavior that creates spatial patterns. To do so, it may be essential to find more common ground on the question of what it means to be human, cultural, and capable of language.

Among many American archaeologists culture has been construed explicitly as humanity's "extrasomatic means of adaptation" (White 1959:8, cited in Binford 1965). Because of the materialist foundation of this theoretical framework, there is a limit to the influence accorded individual intention—most workers prefer to study what they perceive as phenomena determined by the environment, technology, or by the interaction of those limiting factors with the subsistence and settlement system. Thus, subsistence and the material aspects of technology have been pre-eminent foci of inquiry until recently. This theoretical framework has facilitated much useful exploration into hominid evolution, and has yielded important insights into past hominid and human behavior. However, until recently both the motivation for, and the cognitive abilities behind those behaviors have been implicitly modelled, or seen as archaeologically invisible, or as epiphenomenal.

I would argue that investigations of modern human origins employing spatial patterning as a basis for making inferences may yet prove capable of supplying novel insights. However this might only occur if a view of culture is entertained that includes many of the qualities of modern humanness that have been ignored or downplayed in theoretical frameworks that regard human culture as an adaptive system mediating between people and the environment. Indeed, a view of culture such as that outlined in Chapter 1, involving a structure of meaning, language, cultural constructs such as age, kinship, gender, and the like—qualities that may be unique to the hominids inhabiting the present-day world—seems to me to hold the most promise for resolving questions of when (modern) humanness emerged. I would agree with many postprocessual archaeologists, who recognize that the social and individual facets of culture say more about humanness than, for example, optimal foraging behavior or tool-assisted subsistence necessarily do, as neither requires a modern human brain.

In view of this, the question of the usefulness of spatial analysis for the study of modern human origins comes down to our ability to recognize not only behaviorally meaningful patterning, but also the meaning-filled cultural constructions behind behavior. In other words, it is not enough simply to recognize patterning in order to reliably identify the emergence of modern humans from the spatial patterning evidence, it must be demonstrated that the patterning could only have come about through the exertion of modern human cognitive abilities, including self-conscious, meaning-laden action on the part of language-using, symbol-manipulating humans. In short, it will be necessary to identify the cognitive abilities behind the behavior that created the patterns.

To their credit, archaeologists are very good at inferring behavior from its archaeological traces. However, as the foregoing research underscores, behavior does not straightforwardly reveal the cognitive faculties behind the behavior—cave bear, pre-modern hominid and modern human behaviors spatially pattern materials in the caves they occupy, yet their motives need not be equivalent. In light of this, the efficacy of any archaeological method for understanding the evolution of modern humans, including spatial analysis, will rely ultimately on the archaeologist's ability to infer the motive for the behavior that created the patterns. Therefore it is important for archaeologists to work at developing ways of interpreting the cognitive faculties behind behavior resulting in spatial patterning, because those cognitive abilities are the key to tracking the evolution and emergence of modern humans. Archaeologists must therefore address the question of what it means to be human, and how humanness can be recognized in the archaeological record. In the process it may be necessary to entertain a concept of culture that is somewhat different than those that have held sway until recently.

In a related vein, it is my belief that, to date, efforts to infer the cognitive abilities of pre-modern hominids have been susceptible to an epistemological flaw. Presumed similarities between modern human and pre-modern hominids based on stone technology can be seen as the result of a misplaced formal analogy. As discussed in Chapter 1, certain pre-modern hominid behaviors that have until now been seen to imply modern human cognitive abilities, such as handaxe production or the Levallois technique of stone tool making, have recently undergone re-examination. As a result, they have been found to be equivocal on the question of the modernity of thought required to produce them. As Dibble's (1987a, 1987b) work implies, and as Davidson and Noble (1993) have argued, the presumed links between modern human and pre-modern hominid cognitive processes are *sufficient* to account for the behavior, but modern human cognitive abilities may not have been *necessary* for the production of Mousterian stone tools, such as handaxes and Levallois flakes. In light of arguments such as these, no assumptions about the cognitive abilities required for Middle Paleolithic stone-tool-making can unquestionably support the conclusion that those stone tools stand as evidence of pre-modern hominid possession of modern human cognitive abilities. Similar arguments could be made regarding putative pre-modern hominid burial, and hunting. As long as there is this doubt regarding the abilities of pre-modern hominids, it will be worthwhile for archaeologists to work at developing new ways of getting at the meaning behind the behaviors.

One of the ways archaeologists can use spatial analysis to improve their inferences—i.e. to elucidate the motives behind behaviors in order to track the evolution of modern human cognition—is to investigate, and broaden, understanding of meaningful cultural constructs that can produce distinctive patterning in the archaeological record. There are several categories of human behavior that occur irrespective of the particular "adaptive" subsistence strategy, and thus may be seen to transcend mere behavior in their playing out. For example, social distinctions along the lines of gender might produce site structures reflecting meaningful categories such as mens' or womens' areas. Provisioning close kin living in close proximity may result in repetitive, interpretable patterns within encampments, as Gargett and Hayden (1991) have argued, and as Enloe has demonstrated for the Paleolithic site of Pincevent (Enloe 1991, 1992; Enloe and David 1989). Unfortunately, such categories have until recently been downplayed in favor of behavior having more to do with resource extraction and movement across the landscape, behaviors that need have no basis in modern human abilities. The challenge, thus, is to aim for a better understanding of the archaeological record.

Finding spatial patterning that reflects modern human and only modern human cognitive abilities is probably a tall order given the paucity of complex patterns discerned at most well-studied recent hunter-gatherer encampments. Add to this the difficulty of recognizing contemporaneity in any unstratified deposits such as may occur in caves, and the problem seems daunting, indeed. In spite of this, I would still recommend employing spatial analysis in efforts to understand the *behavior* of pre-modern and modern hominids. That, at least, seems feasible. However, understanding the motivation for particular behaviors, and especially the humanness embodied by any behavior, must remain a long-term goal.

In the near term, spatial analysts, and indeed all paleo-anthropologists, need to develop knowledge of the kind of behaviors indicative of meaning-laden cognitive processes. Such investigations may yield insights capable of unlocking the undeveloped potential in spatial analysis for studying modern human origins. If uniquely modern human behaviors can be recognized in the horizontal distributions of bones and artifacts, spatial analysis should ultimately be capable of doing more than simply recognizing *behaviorally* meaningful patterning—it may produce evidence of meaningful *cultural* constructs that are definitive of modern human behavior.

By elucidating meaningful cultural constructions behind behavior, spatial analysis would help to distinguish between hominids that did not have culture and those that did. In summary, when investigating spatial patterning in pre-modern hominid sites, it will not be enough simply to identify patterning that resembles the patterning created by modern humans—it will have to be demonstrated that those patterns are the result of behavior motivated by modern human cognitive abilities. Toward that end, recent postmodern foment in archaeological theory has awakened a sensitivity to the value of research programs involving social categories not explicitly tied to subsistence (e.g. Bender 1978; Conkey and Spector 1984; Gero and Conkey 1991, and papers therein; Hodder 1985; Shanks and Tilley 1987a, 1987b; Wylie 1982). In the end, it may be *just* those social and individual components of culture that have been sublimated in much archaeological theory until recently, and that are the focus of postmodern theorizing in archaeology, that hold the most promise for resolving questions of when humans, like ourselves, emerged.

Conclusion

This research discovered behaviorally meaningful spatial patterning in the Late Pleistocene cave bear bone accumulation at Pod hradem Cave, Czech Republic. The carcasses of animals that had died during hibernation were scavenged by carnivores, predominantly wolves, and the modified bones came to rest in patterned ways. This research has implications for the archaeology of modern human origins. The models that are used to interpret pre-modern hominid behavior from spatial patterns must now include the knowledge that other animals can create spatial patterns, in the absence of culture.

Because of the new ambiguity that this research has introduced, ways must now be sought to get beyond the behavior that creates the patterning to focus instead on modelling the cognitive processes that motivated the behavior. Effort must be made to identify, if possible, distinctive spatial patterns resulting from cognitive processes that are unique to modern humans, i.e. linguistic, meaning making, cultural beings. Only by understanding the cognitive processes behind the behavior that created spatial patterning found at pre-modern hominid archaeological sites will spatial analysis achieve its potential as a means to help resolve questions about the origins of modern humans.

APPENDIX 1

The following pages contain a complete faunal inventory for Pod hradem Cave Stratum 17.

Table A1.1. Pod hradem Cave Stratum 17: Artiodactyl indeterminate

ELEMENT	LEFT	AXIAL or INDET	RIGHT	TOTAL
HORNCORE OR ANTLER		6		6
TOOTH INDETERMINATE		1		1
LUMBAR		1		1
RADIUS, PROXIMAL		1		1
RADIUS, DISTAL		1		1
PROXIMAL PHALANX		2		2
METAPODIAL, SHAFT		6		6
TOTAL		**18**		**18**

Table A1.2. Pod hradem Cave Stratum 17: *Rangifer tarandus*

ELEMENT	LEFT	AXIAL or INDET	RIGHT	TOTAL
TOOTH INDETERMINATE		1		1
SCAPULA		1		1
FEMUR, SHAFT		1		1
ASTRAGALUS		2		2
TOTAL		**5**		**5**

Table A1.3. Pod hradem Cave Stratum 17: Bovid indeterminate

ELEMENT	LEFT	AXIAL or INDET	RIGHT	TOTAL
PROXIMAL PHALANX		1		1
TOTAL		**1**		**1**

Table A1.4. Pod hradem Cave Stratum 17: *Rupicapra rupicapra*

ELEMENT	LEFT	AXIAL or INDET	RIGHT	TOTAL
CRANIUM		1		1
TOOTH				
INDETERMINATE		4		4
TOTAL		**5**		**5**

Table A1.5. Pod hradem Cave Stratum 17: *Capra ibex*

ELEMENT	LEFT	AXIAL or INDET	RIGHT	TOTAL
HORNCORE OR ANTLER		1		1
TOTAL		**1**		**1**

Table A1.6. Pod hradem Cave Stratum 17: Carnivore indeterminate

ELEMENT	LEFT	AXIAL or INDET	RIGHT	TOTAL
CRANIUM		1		1
TOOTH INDETERMINATE		3		3
ULNA, PROXIMAL	1			1
TOTAL	**1**	**4**		**5**

Table A1.7. Pod hradem Cave Stratum 17: *Ursus spelaeus*

ELEMENT	LEFT	AXIAL or INDET	RIGHT	TOTAL
CRANIUM	60	21	71	152
MANDIBLE	46	21	55	122
UPPER TOOTH	30	14	42	86
LOWER TOOTH	41	3	60	104
TOOTH INDET		24		24
HYOID		18		18
ATLAS		9		9
AXIS		9		9
OTHER CERVICAL		23		23
THORACIC		95		95
LUMBAR		37		37
SACRUM		7		7
CAUDAL		3		3
VERTEBRAE INDET	1	16		17
SCAPULA	7	15	11	33
HUMERUS, COMPLETE	3		1	4
HUMERUS, PROXIMAL		1		1
HUMERUS, SHAFT	19	8	29	56
HUMERUS, DISTAL			1	1
RADIUS, COMPLETE			1	1
RADIUS, PROXIMAL	1		3	4
RADIUS, SHAFT	25	1	18	44
RADIUS, DISTAL	1			1
ULNA, COMPLETE			1	1
ULNA, PROXIMAL	1		2	3
ULNA, SHAFT	16	1	21	38
ULNA, DISTAL	1		1	2
CARPALS, COMPLETE	10		5	15
METACARPAL, COMPLETE	11		14	25
METACARPAL, PROXIMAL			3	3
PROXIMAL PHALANX		21		21
MIDDLE PHALANX		7		7
TERMINAL PHALANX		17		17
SESAMOID		1		1
STERNUM		6		6
RIB		202		202
INNOMINATE	8	1	12	21
BACULUM		17		17
FEMUR, COMPLETE	2		2	4
FEMUR, PROXIMAL	1	2	4	7
FEMUR, SHAFT	33	9	40	82
FEMUR, DISTAL	1		2	3

Table A1.7, continued.

ELEMENT	LEFT	AXIAL or INDET	RIGHT	TOTAL
PATELLA	5	2	7	14
TIBIA, COMPLETE	3		2	5
TIBIA, SHAFT	8	1	21	30
TIBIA, DISTAL	4		1	5
FIBULA, PROXIMAL		1	1	2
FIBULA, SHAFT		24		24
FIBULA, DISTAL	3		3	6
CALCANEUS	6		7	13
ASTRAGALUS	3		3	6
OTHER TARSALS			3	3
METATARSAL, COMPLETE	11		15	26
METATARSAL, PROXIMAL	2		1	3
PODIAL		1		1
METAPODIAL, PROXIMAL		4		4
METAPODIAL, SHAFT		13		13
TOTAL	363	655	463	1481

Table A1.8. Pod hradem Cave Stratum 17: *Vulpes vulpes*

ELEMENT	LEFT	AXIAL or INDET	RIGHT	TOTAL
HUMERUS, DISTAL		1		1
TOTAL		1		1

Table A1.9. Pod hradem Cave Stratum 17: *Canis lupus*

ELEMENT	LEFT	AXIAL or INDET	RIGHT	TOTAL
HUMERUS, PROXIMAL		1		1
ULNA, PROXIMAL		1		1
PROXIMAL PHALANX		1		1
FEMUR, DISTAL		1		1
METAPODIAL, COMPLETE		2		2
METAPODIAL, PROXIMAL		1		1
TOTAL		7		7

Table A1.10. Pod hradem Cave Stratum 17: *Crocuta* sp. indet.

ELEMENT	LEFT	AXIAL or INDET	RIGHT	TOTAL
TOOTH INDETERMINATE		1		1
TOTAL		1		1

Table A1.11. Pod hradem Cave Stratum 17: Small mammal

ELEMENT	LEFT	AXIAL or INDET	RIGHT	TOTAL
CRANIUM		1		1
TIBIA, COMPLETE		1		1
FLAT BONE		1		1
TOTAL		**3**		**3**

Table A1.12. Pod hradem Cave Stratum 17: Small/medium mammal

ELEMENT	LEFT	AXIAL or INDET	RIGHT	TOTAL
RIB		2		2
TOTAL		**2**		**2**

Table A1.13. Pod hradem Cave Stratum 17: Large mammal, not *Ursus spelaeus*

ELEMENT	LEFT	AXIAL or INDET	RIGHT	TOTAL
SCAPULA		2		2
CALCANEUS		1		1
UNIDENTIFIABLE		3		3
LONG BONE		6		6
TOTAL		**12**		**12**

Table A1.14. Pod hradem Cave Stratum 17: Large mammal

ELEMENT	LEFT	AXIAL or INDET	RIGHT	TOTAL
CRANIUM		104		104
MANDIBLE		22		22
OTHER CERVICAL		2		2
THORACIC		3		3
VERTEBRAE, INDET		52		52
SCAPULA		57		57
RIB		316		316
INNOMINATE		5		5
FEMUR, SHAFT			2	2
UNIDENTIFIABLE		48		48
EPIPHYSIS, INDET		13		13
LONG BONE		133		133
FLAT BONE		59		59
TOTAL		**814**	**2**	**816**

Table A1.15. Pod hradem Cave Stratum 17: Very large mammal, not *Ursus spelaeus*

ELEMENT	LEFT	AXIAL or INDET	RIGHT	TOTAL
METAPODIAL, SHAFT		1		1
LONG BONE		1		1
TOTAL		**2**		**2**

Table A1.16. Pod hradem Cave Stratum 17: Mammal, size indeterminate

ELEMENT	LEFT	AXIAL or INDET	RIGHT	TOTAL
CRANIUM		5		5
VERTEBRAE, INDET		1		1
UNIDENTIFIABLE		29		29
EPIPHYSIS, INDET		2		2
FLAT BONE		1		1
TOTAL		**37**		**37**

Table A1.17. Pod hradem Cave Stratum 17: Medium bird

ELEMENT	LEFT	AXIAL or INDET	RIGHT	TOTAL
ULNA, PROXIMAL		1		1
TOTAL		**1**		**1**

Table A1.18. Pod hradem Cave Stratum 17: Medium/large bird

ELEMENT	LEFT	AXIAL or INDET	RIGHT	TOTAL
STERNUM		1		1
TOTAL		**1**		**1**

APPENDIX 2

This appendix contains the data used in the spatial analysis of the Pod hradem Cave fauna. Because the horizontal provenience for a given arbitrary level sometimes encompassed more than one meter square unit, each of the following tables displays (1) the raw measures for each unit as it was excavated, (2) a hypothetical total for each individual meter square unit based on the average over the number of units in the originally excavated unit, and (3) where only partial meter squares were excavated, a hypothetical total, which is an estimate of the number of items or the total measure for a complete meter square unit, given the number or the measure for a partial square of known size (see Figure 3.3 for locations of individual meter squares).

Table A2.1. Pod hradem Cave Stratum 17: data accompanying Figure 7.1, horizontal distribution of all specimens (n=2402). Improbably high counts denoted by a "+"; improbably low, by a "-".

UNIT	COUNT[1]	AVERAGE[2]	EXTRAPOLATED[3]	P[4]
31	54	196	-	+0.0000
31,32	273	N/A	N/A	N/A
31,33	9	N/A	N/A	N/A
32	1	148	-	+0.0000
32,34	21	N/A	N/A	N/A
33	0	4	-	-0.0000
34	0	10	-	-0.0000
35	31	31	41	-0.0015
36	32	32	192	+0.0000
37	0	4	-	-0.0000
37,38	8	N/A	N/A	N/A
38	32	53	-	0.0321
38,41	34	N/A	N/A	N/A
39	17	17	-	-0.0000
40	0	0	-	-0.0000
41	28	45	-	0.0059
42	35	35	-	-0.0001
43	13	22	-	-0.0000
43,44	18	N/A	N/A	N/A
44	53	68	-	0.0324
44-46;48-50	34	N/A	N/A	N/A
45	97	117	-	+0.0000
45,46	28	N/A	N/A	N/A
46	40	60	120	+0.0000
47	24	30	-	-0.0000
47,48	10	N/A	N/A	N/A
48	117	135	-	+0.0000
49	83	101	-	+0.0000
49,50	31	N/A	N/A	N/A
50	58	114	-	+0.0000
50,55	10	N/A	N/A	N/A
50,56	31	N/A	N/A	N/A
51	33	33	99	+0.0000
52	1	N/A	N/A	N/A
53	7	N/A	N/A	N/A
54	2	N/A	N/A	N/A
57	3	N/A	N/A	N/A
59	15	N/A	N/A	N/A
61	50	53	-	0.0321
62	39	39	-	-0.0007
63	53	72	-	0.0178

Table A2.1, continued.

UNIT	COUNT[1]	AVERAGE[2]	EXTRAPOLATED[3]	P[4]
63,64	8	N/A	N/A	N/A
64	3	7	-	-0.0000
65	29	29	-	-0.0000
66	66	66	-	0.0400
67	57	57	-	0.0469
68	27	27	-	-0.0000
69	74	74	-	0.0121
70	51	51	-	0.0238
71	5	5	-	-0.0000
72	45	61	-	0.0515
72,73	32	N/A	N/A	N/A
73	0	16	80	+0.0028
74	52	52	-	0.0279
75	54	54	-	0.0363
76	10	45	-	0.0060
76,77	70	N/A	N/A	N/A
77	7	42	84	+0.0008
78	22	42	-	-0.0015
78,82	39	N/A	N/A	N/A
79	71	89	-	+0.0001
79,83	36	N/A	N/A	N/A
80	73	81	108	+0.0000
80,84	16	N/A	N/A	N/A
82	0	19	38	-0.0004
83	9	27	81	+0.0021
84	30	38	228	+0.0000
85	0	0	-	-0.0000
86	30	30	-	-0.0000
87	37	37	-	-0.0003
88	24	24	-	-0.0000
89	0	0	-	-0.0000
90	0	0	-	-0.0000

[1] Count = Actual counts for each provenience unit
[2] Average = Actual counts from provenience units averaged across meter-square grid units
[3] Extrapolated = Averaged counts inflated to what would be expected for an entire meter square
[4] P = exact probablity of occurrence of extrapolated counts, measured against a binomial probability distribution with: number of trials = 2402; number of successes = 228; probability of success = 0.02.

Table A2.2. Pod hradem Cave Stratum 17: data accompanying Figure 7.2, horizontal distribution of average specimen length (mm) per meter square (n=2297). Improbably high average lengths denoted by a "+"; improbably low, by a "-".

METER SQUARE	AVERAGE LENGTH (MM)[1]	P[2]
31	58	0.0405
32	41	-0.0006
33	41	-0.0006
34	89	+0.0004
35	58	0.0405
36	84	+0.0009
37	46	-0.0037
38	70	0.0352
39	135	+0.0000
40	0	-0.0000
41	75	0.0176
42	66	0.0475
43	65	0.0492
44	70	0.0176
45	75	0.0178
46	85	+0.0016
47	62	0.0500
48	63	0.0505
49	67	0.0451
50	72	0.0278
51	88	+0.0006
61	60	0.0465
62	76	0.0147
63	64	0.0503
64	63	0.0505
65	55	0.0289
66	67	0.0451
67	75	0.0176
68	98	+0.0000
69	64	0.0503
70	69	0.0388
71	100	+0.0000
72	73	0.0242
73	59	0.0438
74	52	0.0174
75	54	0.0249
76	68	0.0421
77	59	0.0438
78	70	0.0352
79	70	0.0352
80	67	0.0451

Table A2.2, continued.

METER SQUARE	AVERAGE LENGTH (MM)[1]	P[2]
82	61	0.0486
83	66	0.0475
84	89	+0.0004
85	0	-0.0000
86	61	0.0486
87	70	0.0352
88	63	0.0505
89	0	-0.0000
90	0	-0.0000

[1] Length = average length of specimen per meter square unit, calculated using weighted statistical averages.
[2] P = exact probablity of occurrence of average lengths, measured against a binomial probability distribution with: number of trials = 3183; number of successes = 135; probability of success = 0.02.

Table A2.3. Pod hradem Cave Stratum 17: data accompanying Figure 7.3, horizontal distribution of average specimen areas (cm^2) per meter square (n=1516). Improbably high average areas denoted by a "+"; improbably low, by a "-".

METER SQUARE	AVERAGE AREA (CM^2)[1]	p[2]
31	29	0.0724
32	13	-0.0002
33	60	+0.0000
34	46	+0.0017
35	21	0.0170
36	32	0.0679
37	42	+0.0080
38	40	0.0152
39	115	+0.000
40	0	-0.0000
41	42	+0.0080
42	27	0.0637
43	21	0.0170
44	24	0.0397
45	45	+0.0026
46	43	+0.0056
47	20	0.0117
48	24	0.0397
49	22	0.0236
50	29	0.0724
51	44	+0.0038
61	22	0.0313
62	40	0.0152
63	40	0.0152
64	14	-0.0004
65	14	-0.0004
66	26	0.0566
67	25	0.0483
68	80	+0.0000
69	20	0.0117
70	30	0.0732
71	59	+0.0000
72	35	0.0479
73	32	0.0679
74	14	-0.0004
75	17	-0.0028
76	34	0.0554
77	22	0.0236
78	29	0.0724
79	27	0.0637
80	34	0.0554

Table A2.3, continued.

METER SQUARE	AVERAGE AREA (CM^2)[1]	P[2]
82	20	0.0117
83	20	0.0117
84	61	+0.0000
85	0	-0.0000
86	16	-0.0016
87	23	0.0313
88	24	0.0397
89	0	-0.0000
90	0	-0.0000

[1] Length = average area of specimen per meter square unit, calculated using weighted statistical averages.
[2] P = exact probablity of occurrence of average areas, measured against a binomial probability distribution with: number of trials = 1516; number of successes = 115; probability of success = 0.02.

Table A2.4. Pod hradem Cave Stratum 17: data accompanying Figure 7.4, horizontal distribution of average specimen weight (gm) per meter square (n=1800). Improbably high weights denoted by a "+"; improbably low, by a "-".

METER SQUARE	AVERAGE WEIGHT (GM)[1]	P[2]
31	30	0.0427
32	12	-0.0000
33	100	+0.0000
34	56	+0.0004
35	14	-0.0000
36	31	0.0497
37	67	+0.0000
38	42	0.0383
39	188	+0.0000
40	0	-0.0000
41	47	0.0123
42	27	0.0220
43	16	-0.0001
44	22	-0.0034
45	76	+0.0000
46	47	0.0123
47	17	-0.0002
48	20	-0.0012
49	31	0.0497
50	30	0.0427
51	50	+0.0050
61	23	0.0054
62	47	0.0123
63	43	0.0320
64	11	-0.0000
65	12	-0.0000
66	30	0.0427
67	22	0.0034
68	123	+0.0000
69	16	-0.0001
70	28	0.0284
71	55	+0.0007
72	35	0.0670
73	35	0.0670
74	8	-0.0000
75	14	-0.0000
76	42	0.0383
77	24	0.0081
78	24	0.0081
79	30	0.0427
80	37	0.0652

Table A2.4, continued.

METER SQUARE	AVERAGE WEIGHT (GM)[1]	P[2]
82	17	-0.0002
83	84	+0.0000
84	63	+0.0000
85	0	-0.0000
86	11	-0.0000
87	20	-0.0012
88	23	0.0054
89	0	-0.0000
90	0	-0.0000

[1] Length = average weight of specimen per meter square unit, calculated using weighted statistical averages.

[2] P = exact probablity of occurrence of average weights, measured against a binomial probability distribution with: number of trials = 1800; number of successes = 188; probability of success = 0.02.

Table A2.5. Pod hradem Cave Stratum 17: data accompanying Figure 7.5, horizontal distribution of adult *Ursus spelaeus* specimens (n=894). Improbably high counts denoted by a "+"; improbably low, by a "-".

UNIT	COUNT[1]	AVERAGE[2]	EXTRAPOLATED[3]	P[4]
31	33	33	-	+0.0004
31,32	32	N/A	N/A	N/A
31,33	9	N/A	N/A	N/A
32	1	23	-	0.0423
32,34	12	N/A	N/A	N/A
33	0	4	-	-0.0001
34	0	0	-	-0.0000
35	12	12	16	0.0902
36	13	13	78	+0.0000
37	0	3	-	-0.0000
37,38	7	N/A	N/A	N/A
38	12	37	-	+0.0000
38,41	21	N/A	N/A	N/A
39	12	12	-	0.0378
40	0	0	-	-0.0000
41	15	35	-	+0.0001
42	11	11	-	0.0252
43	8	11	-	0.0252
43,44	7	N/A	N/A	N/A
44	16	26	-	0.0152
44-46;48-50	19	N/A	N/A	N/A
45	54	60	-	+0.0000
45,46	11	N/A	N/A	N/A
46	30	35	70	+0.0000
47	11	11	-	0.0252
47,48	1	N/A	N/A	N/A
48	50	52	-	+0.0000
49	30	37	-	+0.0000
49,50	13	N/A	N/A	N/A
50	19	35	-	+0.0001
50,55	2	N/A	N/A	N/A
50,56	18	N/A	N/A	N/A
51	15	15	45	+0.0000
53	1	N/A	N/A	N/A
57	3	N/A	N/A	N/A
59	4	N/A	N/A	N/A
61	20	23	-	0.0423
62	62	19	-	0.0889
63	27	33	-	+0.0004
63,64	3	N/A	N/A	N/A
64	0	1	-	-0.0000

Table A2.5, continued.

UNIT	COUNT[1]	AVERAGE[2]	EXTRAPOLATED[3]	P[4]
65	2	2	-	-0.0000
66	19	19	-	0.0889
67	24	24	-	0.0313
68	16	16	-	0.0902
69	20	20	-	0.0794
70	23	23	-	0.0423
71	3	3	-	-0.0000
72	14	23	-	0.0423
72,73	17	N/A	N/A	N/A
73	0	8	40	+0.0000
74	9	9	-	0.0085
75	20	20	-	0.0794
76	6	14	-	0.0672
76,77	16	N/A	N/A	N/A
77	4	12	24	0.0313
78	10	19	-	0.0889
78,82	18	N/A	N/A	N/A
79	30	36	-	+0.0001
79,83	12	N/A	N/A	N/A
80	22	27	36	+0.0001
80,84	10	N/A	N/A	N/A
82	0	9	18	0.0945
83	2	8	24	0.0313
84	16	21	126	+0.0000
85	0	0	0	-0.0000
86	11	11	-	0.0252
87	20	20	-	0.0794
88	9	9	-	0.0085
89	0	0	-	-0.0000
90	0	0	-	-0.0000

[1] Count = Actual counts for each provenience unit
[2] Average = Actual counts from provenience units averaged across meter-square grid units
[3] Extrapolated = Averaged counts inflated to what would be expected for an entire meter square
[4] P = exact probablity of occurrence of extrapolated counts, measured against a binomial probability distribution with: number of trials = 894; number of successes = 126; probability of success = 0.02.

Table A2.6. Pod hradem Cave Stratum 17: data accompanying Figure 7.6, horizontal distribution of subadult *Ursus spelaeus* specimens (n=271). Improbably high counts denoted by a "+"; improbably low, by a "-".

UNIT	COUNT[1]	AVERAGE[2]	EXTRAPOLATED[3]	P[4]
31	8	14	-	0.0084
31,32	12	N/A	N/A	N/A
31,33	0	N/A	N/A	N/A
32	0	7	-	0.1499
32,34	2	N/A	N/A	N/A
33	0	0	-	-0.0007
34	0	1	-	0.0050
35	4	4	5	0.1196
36	5	5	30	+0.0000
37	0	0	-	-0.0007
37,38	0	N/A	N/A	N/A
38	2	5	-	0.1196
38,41	6	N/A	N/A	N/A
39	1	1	-	0.0050
40	0	0	-	-0.0007
41	4	7	-	0.1499
42	1	1	-	0.0050
43	0	1	-	0.0050
43,44	1	N/A	N/A	N/A
44	7	8	-	0.1354
44-46;48-50	3	N/A	N/A	N/A
45	9	11	-	0.0507
45,46	3	N/A	N/A	N/A
46	3	5	10	0.0779
47	3	4	-	0.0821
47,48	0	N/A	N/A	N/A
48	18	19	-	+0.0001
49	13	15	-	+0.0040
49,50	1	N/A	N/A	N/A
50	5	10	-	0.0779
50,55	1	N/A	N/A	N/A
50,56	5	N/A	N/A	N/A
51	6	9	26	+0.0000
52	1	N/A	N/A	N/A
53	0	N/A	N/A	N/A
54	1	N/A	N/A	N/A
57	0	N/A	N/A	N/A
59	2	N/A	N/A	N/A
61	4	4	-	0.0821
62	5	5	-	0.1196
63	5	8	-	0.1354

Table A2.6, continued.

UNIT	COUNT[1]	AVERAGE[2]	EXTRAPOLATED[3]	P[4]
63,64	1	N/A	N/A	N/A
64	0	1	-	0.0050
65	6	6	-	0.1449
66	11	11	-	0.0507
67	9	9	-	0.1084
68	3	3	-	0.0450
69	14	14	-	0.0084
70	6	6	-	0.1449
71	2	2	-	0.0184
72	7	8	-	0.1354
72,73	2	N/A	N/A	N/A
73	0	1	5	0.1196
74	7	7	-	0.1499
75	3	3	-	0.0450
76	0	4	-	0.0821
76,77	8	N/A	N/A	N/A
77	1	5	10	0.0779
78	4	7	-	0.1499
78,82	6	N/A	N/A	N/A
79	4	7	-	0.1499
79,83	5	N/A	N/A	N/A
80	9	10	13	0.0165
80,84	1	N/A	N/A	N/A
82	0	3	6	0.1449
83	3	6	17	+0.0007
84	2	3	15	+0.0040
85		0	-	-0.0007
86	6	6	-	0.1449
87	6	6	-	0.1449
88	4	4	-	0.0821
89	0	0	-	-0.0007
90	0	0	-	-0.0007

[1] Count = Actual counts for each provenience unit
[2] Average = Actual counts from provenience units averaged across meter-square grid units, rounded to whole numbers
[3] Extrapolated = Averaged counts inflated to what would be expected for an entire meter square, based on average count per unit
[4] P = exact probablity of occurrence of extrapolated counts, measured against a binomial probability distribution with: number of trials = 271; number of successes = 30; probability of success = 0.02.

Table A2.7. Pod hradem Cave Stratum 17: data accompanying Figure 7.7, horizontal distribution of fetal/neonate *Ursus spelaeus* specimens (n=281). Improbably high counts denoted by a "+"; improbably low, by a "-".

UNIT	COUNT[1]	AVERAGE[2]	EXTRAPOLATED[3]	P[4]
31	1	13	-	0.0168
31,32	23	N/A	N/A	N/A
31,33	0	N/A	N/A	N/A
32	0	12	-	0.0306
32,34	1	N/A	N/A	N/A
33	0	0	-	-0.0007
34	0	0	-	-0.0007
35	8	8	-	0.1357
36	3	3	18	+0.0003
37	0	0	-	-0.0007
37,38	0	N/A	N/A	N/A
38	4	5	-	0.1189
38,41	2	N/A	N/A	N/A
39	2	2	-	0.0181
40	0	0	-	-0.0007
41	1	2	-	0.0181
42	3	3	-	0.0444
43	1	1	-	-0.0049
43,44	1	N/A	N/A	N/A
44	8	9	-	0.1089
44-46;48-50	0	N/A	N/A	N/A
45	9	12	-	0.0306
45,46	6	N/A	N/A	N/A
46	2	5	10	0.0785
47	2	2	-	0.0181
47,48	1	N/A	N/A	N/A
48	10	11	-	0.0513
49	12	15	-	+0.0040
49,50	6	N/A	N/A	N/A
50	7	11	-	0.0513
50,55	0	N/A	N/A	N/A
50,56	1	N/A	N/A	N/A
51	2	2	6	0.1444
53	0	N/A	N/A	N/A
57	0	N/A	N/A	N/A
59	3	N/A	N/A	N/A
61	7	7	-	0.1499
62	5	5	-	0.1189
63	3	7	-	0.1499
63,64	1	N/A	N/A	N/A
64	0	0	-	-0.0007

Table A2.7, continued.

UNIT	COUNT[1]	AVERAGE[2]	EXTRAPOLATED[3]	P[4]
65	5	5	-	0.1189
66	7	7	-	0.1499
67	7	7	-	0.1499
68	1	1	-	-0.0049
69	12	12	-	0.0306
70	4	4	-	0.0814
71	0	0	-	-0.0007
72	7	11		0.0513
72,73	7	N/A	N/A	N/A
73	0	3	15	+0.0040
74	15	15	-	+0.0040
75	15	15	-	+0.0040
76	1	8	-	0.1357
76,77	14	N/A	N/A	N/A
77	0	7	14	0.0086
78	1	6	-	0.1444
78,82	9	N/A	N/A	N/A
79	13	13	-	0.0168
79,83	6	N/A	N/A	N/A
80	10	11	15	+0.0040
80,84	2	N/A	N/A	N/A
82	0	4	8	0.1357
83	0	0	0	-0.0007
84	5	6	36	+0.0000
85	0	0	-	-0.0007
86	4	4	-	0.0814
87	4	4	-	0.0814
88	3	3	-	0.0444
89	0	0	-	-0.0000
90	0	0	-	-0.0000

[1] Count = Actual counts for each provenience unit
[2] Average = Actual counts from provenience units averaged across meter-square grid units
[3] Extrapolated = Averaged counts inflated to what would be expected for an entire meter square
[4] P = exact probablity of occurrence of extrapolated counts, measured against a binomial probability distribution with: number of trials = 362; number of successes = 36; probability of success = 0.02.

References Cited

Abel, O., and G. Kyrle, eds. 1931. *Die Drachenhöhle bei Mixnitz.* Speläologische Monographien 7-8. Vienna: Speläologisches Institut.

Agenbroad, Larry D. 1989. Spiral fractured mammoth bone from nonhuman taphonomic processes at Hot Springs Mammoth site. In *Bone modification*, ed. Robson Bonnichsen and Marcella H. Sorg, 139-148. Orono, Maine: Center for the Study of the First Americans, University of Maine.

Akazawa, Takeru, Kinichi Aoki, and Tasuku Kimura, eds. 1992. *The evolution and dispersal of modern humans in Asia*. Tokyo: Hokusen-Sha.

Andrews, Peter. 1990. *Owls, caves and fossils*. Chicago: University of Chicago Press.

Andrews, Peter, and Jill Cook. 1985. Natural modifications to bones in a temperate setting. *Man* (n.s.) 20: 675-91.

Bächler, E. 1921. Das Drachenloch bei Vättis im Taminatal. *Jahrbuches der St. Gallischen Naturwissenschaftlichen Gesellschaft* 57(I).

Badgley, Catherine. 1986. Counting individuals in mammalian fossil assemblages from fluvial environments. *Palaios* 1: 328-38.

Bahn, Paul G. 1983. The case of the clumsy cave-bears. *Nature* 301: 565.

Bar-Yosef, Ofer, B. Vandermeersch, B. Arensburg, A. Belfer-Cohen, P. Goldberg, H. Laville, L. Meignen, Y. Rak, J. D. Speth, E. Tchernov, A.-M. Tillier, and S. Weiner. 1992. The excavations in Kebara Cave, Mt. Carmel. *Current Anthropology* 33: 497-550.

Barton, C. Michael. 1990. Beyond style and function: a view from the Middle Paleolithic. *American Anthropologist* 92: 57-72.

Behrensmeyer, Anna K. 1973. The taphonomy and paleoecology of Plio-Pleistocene vertebrate assemblages east of Lake Rudolph, Kenya. Ph.D. diss., Harvard University.

________. 1975. The taphonomy and paleoecology of Plio-Pleistocene vertebrate assemblages east of Lake Rudolph, Kenya. *Bulletin of the Museum of Comparative Zoology, Harvard University* 146: 473-578.

________. 1978. Taphonomic and ecologic information from bone weathering. *Paleobiology* 4: 150-62.

________. 1982. Time resolution in fluvial vertebrate assemblages. *Paleobiology* 8: 211-27.

Behrensmeyer, Anna K., and Dorothy E. Dechant Boaz. 1980. The recent bones of Amboseli Park, Kenya, in relation to East African Paleoecology. In *Fossils in the making: vertebrate taphonomy and paleoecology*, ed. Anna K. Behrensmeyer and Andrew P. Hill, 72-92. Chicago: University of Chicago Press.

Behrensmeyer, Anna K., Kathleen D. Gordon, and Glenn T. Yanagai. 1986. Trampling as a cause of bone surface damage and pseudo-cutmarks. *Nature* 319: 768-71.

________. 1989. Nonhuman bone modification in Miocene fossils from Pakistan. In *Bone modification*, ed. Robson Bonnichsen and Marcella H. Sorg, 99-120. Orono, Maine: Center for the Study of the First Americans, University of Maine.

Behrensmeyer, Anna K., and Andrew P. Hill, eds. 1980. *Fossils in the making: vertebrate taphonomy and paleoecology*. Chicago: University of Chicago Press.

Behrensmeyer, Anna K., and Susan M. Kidwell. 1985. Taphonomy's contribution to paleobiology. *Paleobiology* 11: 105-19.

Belfer-Cohen, Anna, and Erella Hovers. 1992. In the eye of the beholder: Mousterian and Natufian burials in the Levant. *Current Anthropology* 33:463-71.

Bender, Barbara. 1978. Gatherer-hunter to farmer: a social perspective. *World Archaeology* 10: 204-22.

Bergounioux, F. M. 1958. Spiritualité de l'homme de Néanderthal. In *Neanderthal centenary*, ed. G. H. R. von Koenigswald, 151-66. New York: Wenner-Gren Foundation.

Binford, Lewis R. 1964. A consideration of archaeological research design. *American Antiquity* 29: 425-41.

________. 1965. Archaeological systematics and the study of culture process. *American Antiquity* 31: 203-10.

________. 1977. General introduction. In *For theory building in archaeology*, ed. Lewis R. Binford, 1-10. New York: Academic Press.

________. 1978a. *Nunamiut ethnoarchaeology*. New York: Academic Press.

________. 1978b. Dimensional analysis of behavior and site structure: learning from an Eskimo hunting stand. *American Antiquity* 43: 330-61.

________. 1981a. Behavioral archaeology and the "Pompeii Premise." *Journal of Anthropological Research* 37: 195-208.

________. 1981b. *Bones: ancient men and modern myths*. New York: Academic Press.

________. 1983. *In pursuit of the past: decoding the archaeological record*. New York: Thames and Hudson.

________. 1984. *Faunal remains from Klasies River Mouth*. New York: Academic Press.

_______. 1985. Human ancestors: Changing views of their behavior. *Journal of Anthropological Archaeology* 4: 292-327.

Binford, Lewis R., and Jack B. Bertram. 1977. Bone frequencies—and attritional processes. In *For theory building in archaeology: essays on faunal remains, aquatic resources, spatial analysis, and systemic modelling*, ed. Lewis R. Binford, 77-153. New York: Academic Press.

Binford, Lewis R., and Sally R. Binford. 1966. A preliminary analysis of functional variability in the Mousterian of Levallois facies. *American Anthropologist* 68: 238-95.

Binford, Lewis R., S. R. Binford, Robert Whallon, and Margaret Ann Hardin. 1970. *Archaeology at Hatchery West*. Memoirs of the Society for American Archaeology 24 [*American Antiquity* 35(4)].

Binford, Lewis R., and Jeremy A. Sabloff. 1982. Paradigms, systematics, and archaeology. *Journal of Anthropological Research* 38: 137-53.

Binford, Sally R. 1971. The significance of variability: a minority report. In *The origin of* Homo sapiens, ed. F. Bordes, 199-210. New York: Unesco.

Binford, Sally R., and Lewis R. Binford. 1968. *New perspectives in archaeology*. Chicago: Aldine.

_______. 1969. Stone tools and human behavior. *Scientific American* 220: 70-84.

Blalock, Hubert M., Jr. 1979. *Social statistics*. Revised 2d ed. New York: McGraw-Hill.

Blumenschine, Robert J. 1986. Carcass consumption sequences and the archaeological distinction of scavenging and hunting. *Journal of Human Evolution* 15: 639-59.

_______. 1987. Characteristics of an early hominid scavenging niche. *Current Anthropology* 28: 383-407.

_______. 1988. An experimental model of the timing of hominid and carnivore influence on archaeological bone assemblages. *Journal of Archaeological Science* 15: 483-502.

_______. 1989. A landscape taphonomic model of the scale of prehistoric scavenging opportunities. *Journal of Human Evolution* 18: 345-71.

_______. 1991. Prey size and age models of prehistoric hominid scavenging: test cases from the Serengeti. In *Human predators and prey mortality*, ed. Mary C. Stiner, 121-47. Boulder, Colorado: Westview Press.

Bocherens, H., M. Fizet, and A. Mariotti. 1990. Mise en évidence du régime alimentaire végétarien de l'ours des cavernes (*Ursus spelaeus*) par la biogéochimie isotopique (^{13}C, ^{15}N) du collagène des Vertébrés fossiles. *Comptes Rendus de l'Academie des Sciences de Paris, Série II* 311: 1279-84.

Bonifay, Eugène. 1962. Un ensemble rituel moustérien à la grotte de Regourdou (Montignac, Dordogne). *Proceedings of the 6th International Congress of Prehistoric and Protohistoric Sciences, Rome*, volume 2, 136-40. Firenze: G. N. Sansoni.

_______. 1964. La grotte de Regourdou (Montignac, Dordogne): stratigraphie et industrie lithic moustérienne. *L'Anthropologie* 68: 49-64.

Bonnichsen, Robson, and Marcella H. Sorg, eds. 1989. *Bone modification.* Orono, Maine: Center for the Study of the First Americans, University of Maine.

Bordes, François. 1953. Essai de classification des industries Mousteriennes. *Bulletin de la Société Préhistorique Française* 50: 457-66.

________. 1961a. Mousterian cultures in France. *Science* 134: 803-10.

________. 1961b. *Typologie du Paléolithique Ancien et Moyen*. Paris: Centre National de Recherche Scientifique.

________. 1980. Le débitage Levallois et ses variants. *Bulletin de la Société Préhistorique Française* 77: 45-9.

Borrero, Luis Alberto. 1990. Taphonomy of guanaco bones in Tierra del Fuego. *Quaternary Research* 34: 361-71.

Bouyssonie, A., J. Bouyssonie, and L. Bardon. 1908. Découverte d'un squelette humain moustérien à la Bouffia de la Chapelle-aux-Saints (Corrèze). *L'Anthropologie* 19: 513-18.

Brain, C. K. 1967. Hottentot food remains and their bearing on the interpretation of fossil bone assemblages. *Scientific Papers of the Namib Research Station* 32.

________. 1968. Who killed the Swartkrans ape-man? *South African Museum Bulletin* 9: 127-39.

________. 1969. The contribution of the Namib Desert Hottentots to an understanding of australopithecine bone accumulations. *Scientific Papers of the Namib Desert Research Station* No. 39.

________. 1974. Some suggested procedures in the analysis of bone accumulations from Southern African Quaternary sites. *Transvaal Museum, Pretoria. Annals* 29: 1-8.

________. 1980. Some criteria for the recognition of bone-collecting agencies in African caves. In *Fossils in the making: vertebrate taphonomy and paleoecology*, ed. Anna K. Behrensmeyer and Andrew P. Hill, 107-30. Chicago: University of Chicago Press.

________. 1981. *The hunters or the hunted? An introduction to African cave taphonomy*. Chicago: University of Chicago Press.

Brain, C. K., and G. Turner. 1984. Problems in unravelling multi-agent involvement in bone accumulations. In *Frontiers: Southern African archaeology today*, ed. M. Hall, G. Avery, D. M. Avery, M. L. Wilson, and A. J. B. Humphreys, 340-43. B.A.R. International Series No. 207. Oxford: British Archaeological Reports.

Bräuer, G. 1984. A craniological approach to the origin of anatomically modern *Homo sapiens* in Africa and implications for the appearance of modern Europeans. In *The origins of modern humans: a world survey of the fossil evidence*, ed. F. H. Smith and F. Spencer, 327-410. New York: Alan R. Liss.

Brew, John O. 1946. The use and abuse of taxonomy. In *Archaeology of Alkalai Ridge, Southeastern Utah, with a review of the prehistory of the Mesa Verde Division of the San Juan and some observations on archaeological systematics*, 44-66. Papers of the Peabody Museum of American Archaeology and Ethnology 21. Cambridge, Massachusetts: Harvard University Press.

Brothwell, Don. 1976. Further evidence of bone chewing by ungulates: the sheep of North Ronaldsay, Orkney. *Journal of Archaeological Science* 3: 179-82.

Brugal, Jean-Philip, and Jacques Jaubert. 1991. Les gisements paléontologiques Pléistocènes à indices de fréquentation humaine: un nouveau type de comportement de prédation? *Paléo* 3: 15-41.

Bunn, Henry. 1983. Comparative analysis of modern bone assemblages from a San hunter-gatherer camp in the Kalahari Desert, Botswana, and from a spotted hyena den near Nairobi, Kenya. In *Animals and archaeology: 1. Hunters and their prey*, ed. Juliet Clutton-Brock and Caroline Grigson, 143-48. B.A.R. International Series 163. Oxford: British Archaeological Reports.

Bunn, Henry T., Laurence E. Bartram, and Ellen M. Kroll. 1988. Variability in bone assemblage formation from Hadza hunting, scavenging, and carcass processing. *Journal of Anthropological Archaeology* 7: 412-57.

Butzer, Karl. 1981. Cave sediments, Upper Pleistocene stratigraphy, and Mousterian facies in Cantabrian Spain. *Journal of Archaeological Science* 8: 133-83.

________. 1982. *Archaeology as human ecology*. New York: Cambridge University Press.

Cann, R. L., M. Stoneking, and A. Wilson. 1987. Mitochondrial DNA and human evolution. *Nature* 325: 31-36.

Carr, Christopher. 1984. The nature of organization of intrasite archaeological records and spatial analytical approaches to their investigation. In *Advances in Archaeological Method and Theory*, vol. 7, ed. M. B. Schiffer, 103-222. New York: Academic Press.

Chase, Philip G. 1991. Symbols and Paleolithic artifacts: style, standardization, and the imposition of arbitrary form. *Journal of Anthropological Archaeology* 10: 193-214.

Chase, Philip G., and Harold Dibble. 1987. Middle Paleolithic symbolism. *Journal of Anthropological Archaeology* 6: 263-96.

Chave, K. E. 1964. Skeletal durability and preservation. In *Approaches to paleoecology*, ed. J. Imbrie and N. Newell, 377-87. New York: Wiley.

Clark, Geoffrey A. 1991. A paradigm is like an onion: reflections on my biases. In *Perspectives on the past*, ed. G. A. Clark, 79-108. Philadelphia: University of Pennsylvania Press.

Clark, Geoffrey A., and John M. Lindly. 1989a. The case for continuity: observations on the biocultural transition in Europe and western Asia. In *The human revolution: behavioural and biological perspectives on the origins of modern humans*, ed. P. Mellars and C. B. Stringer, 626-76. Edinburgh: Edinburgh University Press.

________. 1989b. Modern human origins in the Levant and western Asia. *American Anthropologist* 91: 962-85.

Clarke, David L. 1977. Spatial information in archaeology. In *Spatial archaeology*, ed. David L. Clarke, 1-32. New York: Academic Press.

Collcutt, S. N. 1979. The analysis of Quaternary cave sediments. *World Archaeology* 10: 290-301.

Conkey, Margaret W., and Janet D. Spector. 1984. Archaeology and the study of gender. In *Advances in Archaeological Method and Theory*, vol. 7, ed. M. B. Schiffer, 1-38. New York: Academic Press.

Cowgill, George L. 1993. Distinguished lecture in Archeology: beyond criticizing New Archeology. *American Anthropologist* 95: 551-73.

Cruz-Uribe, Kathryn. 1991. Distinguishing hyena from hominid bone accumulations. *Journal of Field Archaeology* 18: 467-86.

Dart, Raymond A. 1949. The predatory implemental technique of Australopithecus. *American Journal of Physical Anthropology* 7:1-38.

Davidson, Iain. 1991. The archaeology of language origins: a review. *Antiquity* 65:39-48.

Davidson, Iain, and William Noble. 1989. The archaeology of perception: traces of depiction and language. *Current Anthropology* 30:125-55.

________. 1992. Why the first colonisation of the Australian region is the earliest evidence of modern human behavior. *Archaeology in Oceania* 27:135-42.

________. 1993. Tools and language in human evolution. In *Tools, language and cognition in human evolution*, ed. K. Gibson and T. Ingold, 363-88. Cambridge: Cambridge University Press.

Davis, Whitney. 1992. The deconstruction of intentionality in archaeology. *Antiquity* 66:334-47.

Dettwyler, K. A. 1991. Can paleopathology provide evidence for "compassion"? *American Journal of Physical Anthropology* 84:375-84.

Dibble, Harold L. 1984. Interpreting typological variation of Middle Paleolithic Scrapers: function, style, or sequence of reduction? *Journal of Field Archaeology* 11:431-36.

________. 1985a. Technological aspects of flake variation. *American Archeology* 5:236-40.

________. 1985b. Raw-material variation in Levallois flake manufacture. *Current Anthropology* 26:391-93.

________. 1987a. The interpretation of Middle Paleolithic scraper morphology. *American Antiquity* 52:109-17.

________. 1987b. Reduction sequences in the manufacture of Mousterian implements of France. In *The Pleistocene Old World*, ed. Olga Soffer, 33-45. New York: Plenum Press.

Dibble, Harold L., and Paul Mellars, eds. 1992. *The Middle Paleolithic: adaptation, behavior, and variability*. University Museum Monograph 78. University Museum Symposium Series Volume IV. Philadelphia: The University Museum, University of Pennsylvania.

Dibble, Harold L., and Anta Montet-White, eds. 1988. *The Upper Pleistocene prehistory of western Eurasia.* Philadelphia: The University Museum, University of Pennsylvania.

Dibble, Harold L., and Nicolas Rolland. 1992. On assemblage variability in the Middle Paleolithic of western Europe: history, perspectives, and a new synthesis. In *The Middle Paleolithic: adaptation, behavior, and variability,* ed. Harold L. Dibble and Paul Mellars, 1-28. University Museum Monograph 78. University Museum Symposium Series Volume IV. Philadelphia: The University Museum, University of Pennsylvania.

Dodson, Peter, and Diane Wexlar. 1979. Taphonomic investigations of owl pellets. *Paleobiology* 5:275-84.

Dreybrodt, Wolfgang. 1988. *Processes in karst systems: physics, chemistry, and geology.* New York: Springer-Verlag.

Driver, Jonathan C. 1982. Minimum standards for reporting animal bones in salvage archaeology: Southern Alberta as a case study. In *Directions in archaeology: a question of goals,* ed. P. D. Francis and E. C. Poplin, 199-209. Calgary: University of Calgary Archaeological Association.

Duff, Andrew, Geoffrey A. Clark, and Thomas J. Chadderdon. 1992. Symbolism in the Early Palaeolithic: a conceptual odyssey. *Cambridge Archaeological Journal* 2:211-29.

Earle, Timothy. 1985. Pattern recognition and the field archeologist: a brief commentary. *American Archeology* 5:77-78.

Enloe, J. G. 1991. Subsistence organization in the Upper Paleolithic: carcass refitting and food sharing at Pincevent. Unpublished Ph.D. diss., Department of Anthropology, University of New Mexico, Albuquerque.

________. 1992. Le partage de la nourriture à partir des témoins archéologiques: une application ethnoarchéologique. In *Ethnoarchéologie: justification, problèmes, limites.* XIIe Rencontres Internationales d'Archéologie et d'Histoire d'Antibes. Juan-les-Pins: Éditions APDCA.

Enloe, J. G., and Francine David. 1989. Le remontage des os par individus: le partage du renne chez les Magdaléniens de Pincevent (La Grande Paroisse, Seine-et-Marne). *Bulletin de la Société Préhistorique Française* 86(9):275-81.

Erlandson, Jon M. 1984. A case study in faunalturbation: delineating the effects of the burrowing pocket gopher on the distribution of archaeological materials. *American Antiquity* 49:785-90.

Falk, D. 1983. Cerebral cortices of east African early hominids. *Science* 221:1072-74.

Farizy, Catherine. 1990. The transition from Middle to Upper Palaeolithic at Arcy-sur-Cure (Yonne, France): technological, economic, and social aspects. In *The emergence of modern humans: an archaeological perspective,* ed. Paul Mellars, 303-26. New York: Cornell University Press.

Farrand, William R. 1985. Rockshelter and cave sediments. In *Archaeological sediments in context,* ed. Julie K. Stein and William R. Farrand, 21-39. Orono, Maine: Center for the Study of Early Man, University of Maine.

Fiorillo, Anthony R. 1989. An experimental study of trampling: implications for the fossil record. In *Bone modification*, ed. Robson Bonnichsen and Marcella H. Sorg, 61-71. Orono, Maine: Center for the Study of the First Americans, University of Maine.

Foley, R. A. 1987. Hominid species and stone tool assemblages: how are they related? *Antiquity* 61:380-92.

Ford, Derek C., and P. W. Willams. 1989. *Karst geomorhpology and hydrology*. Boston: Unwin Hyman.

Ford, T. D. 1976. The geology of caves. In *The science of speleology*, ed. T. D. Ford and C. H. D. Cullingford, 11-60. New York: Academic Press.

Ford, T. D. and C. H. D. Cullingford, eds. 1976. *The science of speleology*. New York: Academic Press.

Freeman, L. G. 1992. Mousterian facies in space: new data from Morín Level 16. In *The Middle Paleolithic: adaptation, behavior, and variability*, ed. Harold L. Dibble and Paul Mellars, 113-25. University Museum Monograph 78. University Museum Symposium Series Volume IV. Philadelphia: The University Museum, University of Pennsylvania.

Gábori-Csánk, V. 1968. *La station du Paléolithique moyen d'Erd—Hongrie*. Budapest: Akadémiai Kiadó, Maison d'Edition de l'Académie des Sciences de Hongrie.

Gargett, Robert H. 1989a. Grave shortcomings: the evidence for Neandertal burial. *Current Anthropology* 30:157-90.

________. 1989b. Reply to comments on "Grave shortcomings: the evidence for Neandertal burial." *Current Anthropology* 30:326-29.

________. 1990. Comment on "Symbolism and modern human origins," by John M. Lindley and Geoffrey A. Clark. *Current Anthropology* 31:244-45.

________. 1991. Comment on "The faunal remains from Grotta Guattari: a taphonomic perspective," by Mary Stiner, and on "The question of ritual cannibalism at Grotta Guattari," by T. D. White and N. Toth. *Current Anthropology* 32:129-30.

________. 1992. What is archaeology, really? *Kroeber Anthropological Society Papers* 73-74:115-48.

________. 1993. Comment on "Glottogenesis and anatomically modern *Homo sapiens*: the evidence for and implications of a late origin of vocal language," by Richard G. Milo and Duane Quiatt. *Current Anthropology* 34:584-85.

Gargett, Robert H., and Brian Hayden. 1991. Site structure, kinship, and sharing in Aboriginal Australia: implications for archaeology. In *The interpretation of archaeological spatial patterning*, ed. Ellen M. Kroll and T. Douglas Price, 11-32. New York: Plenum Press.

Gero Joan M., and Margaret W. Conkey, eds. 1991. *Engendering archaeology*. Oxford: Blackwell.

Gifford, Diane P. 1977. Observations of modern human settlements as an aid to archaeological interpretation. Ph.D. diss., Department of Anthropology, University of California, Berkeley. Ann Arbor, Michigan: University Microfilms International.

_______. 1981. Taphonomy and paleoecology: a critical review of archaeology's sister disciplines. In *Advances in Archaeological Method and Theory*, vol. 4, ed. M. B. Schiffer, 365-438. New York: Academic Press.

Gifford-Gonzalez, Diane P. 1989. Modern analogues: developing an interpretive framework. In *Bone modification*, ed. Robson Bonnichsen and Marcella H. Sorg, 43-52. Orono, Maine: Center for the Study of the First Americans, University of Maine.

_______. 1991. Bones are not enough: analogues, knowledge, and interpretive strategies in zooarchaeology. *Journal of Anthropological Archaeology* 10:215-54.

_______. 1993. Gaps in the zooarchaeological record. In *From bones to behavior: ethnoarchaeological and experimental contributions to the interpretation of faunal remains*, ed. Jean Hudson, 181-195. Occasional Paper 21. Carbondale, Illinois: Southern Illinois University Press.

Gifford-Gonzalez, Diane P., D. B. Damrosch, D. R. Damrosch, J. Pryor, and R. L. Thunen. 1985. The third dimension in site structure: an experiment in trampling and vertical dispersal. *American Antiquity* 50:803-18.

Glover, Ian C. 1979. The effects of sink action on archaeological deposits in caves: an Indonesian example. *World Archaeology* 10:302-17.

Gordon, Brian C. 1976. Antler pseudo-tools made by caribou. In *Primitive art and technology*, ed. J. S. Raymond, 121-28. Calgary: University of Calgary Archaeological Association.

Gould, Richard J., and John E. Yellen. 1987. Man the hunted: determinants of household spacing in desert and tropical foraging societies. *Journal of Anthropological Archaeology* 6:77-103.

Grant, A. 1982. The use of tooth wear as a guide to the age of domestic animals. In *Aging and sexing animal bones from archaeological sites*, ed. B. Wilson, C. Grigson, and S. Payne, 91-108. B.A.R. British Series 109. Oxford: British Archaeological Reports.

Grayson, Donald K. 1984. *Quantitative zooarchaeology*. Academic Press, New York.

_______. 1988. *Danger Cave, Last Supper Cave, and Hanging Rock Shelter: the faunas*. American Museum of Natural History Anthropological Papers 66(1). New York: American Museum of Natural History.

_______. 1989. Bone transport, bone destruction, and reverse utility curves. *Journal of Archaeological Science* 16:643-52.

Hanson, C. Bruce. 1980. Fluvial taphonomic processes: models and experiments. In *Fossils in the making: vertebrate taphonomy and paleoecology*, ed. Anna K. Behrensmeyer and Andrew P. Hill, 156-81. Chicago: University of Chicago Press.

Harris, Marvin. 1979. *Cultural materialism: the struggle for a science of culture*. New York: Random House.

Harrold, Frank B. 1980. A comparative analysis of Eurasian Paleolithic burials. *World Archaeology* 12:195-211.

Hayden, Brian. 1993. The cultural capacities of Neandertals: a review and re-evaluation. *Journal of Human Evolution* 24:113-46.

Hayden, Brian, and Aubrey Cannon. 1983. Where the garbage goes: refuse disposal in the Maya Highlands. *Journal of Anthropological Archaeology* 2:117-63.

Haynes, Gary. 1980. Evidence of carnivore gnawing on Pleistocene and recent mammalian bone. *Paleobiology* 6:341-51.

________. 1981. Bone modification and skeletal disturbances by natural agencies: studies in North America. Ph.D. diss., Department of Anthropology, Catholic University of America, Washington, D.C. Ann Arbor, Michigan: University Microfilms.

________. 1982. Utilization and skeletal disturbances of North American prey carcasses. *Arctic* 35:266-81.

________. 1983a. Frequencies of spiral and green-bone fractures on ungulate limb bones in modern surface assemblages. *American Antiquity* 48:102-14.

________. 1983b. A guide for differentiating mammalian carnivore taxa responsible for gnaw damage to herbivore limb bones. *Paleobiology* 9:164-72.

________. 1985. Age profiles in elephant and mammoth bone assemblages. *Quaternary Research* 24:333-45.

________. 1988a. Longitudinal studies of African elephant death and bone deposits. *Journal of Archaeological Science* 15:131-57.

________. 1988b. Mass deaths and serial predation: comparative taphonomic studies of large mammal death sites. *Journal of Archaeological Science* 15:219-35.

Henschel, J. R., R. Tilson, and F. von Blottnitz. 1979. Implications of a spotted hyaena bone assemblage in the Namib desert. *South African Archaeological Bulletin* 34:127-31.

Hesse, Brian. 1982. Bias in the zooarchaeological record: suggestions for interpretation of bone counts in faunal samples from the Plains. In *Plains Indian Studies: A collection of essays in honor of John C. Ewers and Waldo R. Wedel*, ed. D. H. Ubelaker and H. J. Viola, 157-72. Smithsonian Contributions in Anthropology 30. Washington, D.C.: Smithsonian Institution.

Hietala, Harold J., and Dominique E. Stevens. 1977. Spatial analysis: multiple procedures in pattern recognition studies. *American Antiquity* 42:539-59.

Hill, Andrew P. 1975. Taphonomy of contemporary and Late Cenozoic East African vertebrates. Ph.D. diss., University of London.

________. 1976. On carnivore damage and weathering damage to bone. *Current Anthropology* 17:335-336.

________. 1977. Disarticulation and scattering of mammal skeletons. *Paleobiology* 5:261-74.

________. 1979. Butchery and natural disarticulation: an investigatory technique. *American Antiquity* 44:739-44.

________. 1980. Early postmortem damage to the remains of some contemporary east African mammals. In *Fossils in the making: vertebrate taphonomy and paleoecology*, ed. Anna K. Behrensmeyer and Andrew P. Hill, 131-55. Chicago: University of Chicago Press.

________. 1983. Hyaenas and early hominids. In *Animals and archaeology: 1. Hunters and their prey*, ed. Juliet Clutton-Brock and Caroline Grigson, 87-92. BAR International Series 163. Oxford: British Archaeological Reports.

________. 1989. Bone modification by modern spotted hyenas. In *Bone modification*, ed. Robson Bonnichsen and Marcella H. Sorg, 169-78. Orono, Maine: Center for the Study of the First Americans, University of Maine.

Hill, James N. 1966. A prehistoric community in Eastern Arizona. *Southwestern Journal of Anthropology* 22:9-30.

Hodder, Ian. 1985. Postprocessual archaeology. In *Advances in Archaeological Method and Theory*, vol. 8, ed. M. B. Schiffer, pp. 1-26. Academic Press, New York.

________. 1986. *Reading the past: current approaches to interpretation in archaeology*. Cambridge: Cambridge University Press.

Holloway, R. L. 1974. The casts of fossil hominid brains. *Scientific American* 231:106-115.

________. 1985. The poor brain of *Homo sapiens neanderthalensis*: see what you please. In *Ancestors: the hard evidence*, ed. E. Delson, 319-24. New York: Alan R. Liss.

Horejsi, B. L., E. Hornbeck, and R. M. Raine. 1984. Wolves, *Canis lupus*, kill female black bear, *Ursus americanus*, in Alberta. *Canadian Field Naturalist* 98:368-69.

Howells, W. W. 1976. Explaining modern man: evolutionists versus migrationists. *Journal of Human Evolution* 5:577-96.

Hughes, A. R. 1961. Further notes on the habits of hyaenas and bone gathering by porcupines. *Zoological Society of South Africa News Bulletin* 3(1).

Jennings, J. N. 1985. *Karst geomorphology*. Oxford: Blackwell.

Johnson, D. L., and C. V. Haynes. 1985. Camels as taphonomic agents. *Quaternary Research* 24:365-66.

Kent, Susan. 1993. Variability in faunal assemblages—the influence of hunting skill, sharing, dogs, and mode of cooking on faunal remains at a sedentary Kalahari community. *Journal of Anthropological Archaeology* 12:323-85.

Kerbis-Peterhans, Julian C. 1990. The roles of porcupines, leopards and hyenas in ungulate carcass dispersal: implications for paleoanthropology. Ph.D. diss., University of Chicago. Ann Arbor, Michigan: University Microfilms.

Kintigh, Keith. 1985-1994. Tools for quantitative archaeology: programs for quantitative analysis in archaeology. Keith W. Kintigh, 2014 East Alameda Drive, Tempe, Arizona 85282-4002 U.S.A.

Klein, Richard G. 1975. Paleoanthropological implications of the nonarchaeological bone assemblage from Swartklip I, Southwestern Cape Province, South Africa. *Quaternary Research* 5:275-88.

________. 1982. Age (mortality) profiles as a means of distinguishing hunted species from scavenged ones in Stone Age archeological sites. *Paleobiology* 8:151-58.

________. 1989. *The human career: human biological and cultural origins*. Chicago: University of Chicago Press.

________. 1992. The archeology of modern human origins. *Evolutionary Anthropology* 1:5-14.

Klein, Richard G., and Kathryn Cruz-Uribe. 1984. *The analysis of animal bones from archeological sites*. Chicago: University of Chicago Press.

Kluckhohn, Clyde. 1939. The place of theory in anthropological studies. *Philosophy of Science* 6:328-44.

Knies, J. 1901. Čtvrtohorní zvířena jeskyně Pod hradem u Suchdola na Moravě. *Časopis Vlasteneckeho Muzejniho spolku v Olomouci* 69:1-15.

Koby, F. Edouard. 1941. Le «Charriage à sec» des ossements dans les cavernes. *Eclogae Geologicae Helvetiae* 34(2):319-20.

________. 1953. Les paléolithiques ont-ils chassé l'ours des cavernes? *Actes de la Société Jurassienne d'Emulation* Series II 57:157-204.

Kroll, Ellen M., and Glynn Ll. Isaac. 1984. Configurations of artifacts and bones at early Pleistocene sites in East Africa. In *Intrasite spatial analysis in archaeology*, ed. Harold J. Hietala, 4-31. Cambridge: Cambridge University Press.

Kroll, Ellen M., and T. Douglas Price, eds. 1991. *The interpretation of archaeological spatial patterning*. New York: Plenum Press.

Kruuk, Hans. 1972. *The spotted hyena: a study of predation and social behavior*. Chicago: University of Chicago Press.

Kukla, Jiří. 1968. Comment on "Evolution of the Palaeolithic in Central and Eastern Europe," by Karel Valoch. *Current Anthropology* 9:378-80.

Kurtén, Bjorn. 1958. Life and death of the Pleistocene cave bear: a study in paleoecology. *Acta Zoologica Fennica* 95.

________. 1976. *The cave bear story*. New York: Columbia University Press.

Lam, Yin M. 1992. Variability in the behaviour of spotted hyaenas as taphonomic agents. *Journal of Archaeological Science* 19:389-406.

Laville, Henri, Jean-Phillipe Rigaud, and James R. Sackett. 1980. *Rock shelters of the Perigord*. New York: Academic Press.

Leach, Edmund. 1976. *Culture and communication: the logic by which symbols are connected*. Cambridge: Cambridge University Press.

Leroi-Gourhan, André. 1950. *Les fouilles préhistoriques (Technique et Méthodes)*. Avec un appendice par Annette Laming. Paris: Picard.

Leroi-Gourhan, André, and M. Brézillon. 1966. L'habitation magdalénienne No. 1 de Pincevent près Montereau (Seine-et-Marne). *Gallia Préhistoire* 2:263-385.

_______. 1972. *Fouilles de Pincevent, essai d'analyse ethnographique d'un habitat Magdalénien (la section 36) (Gallia Préhistoire, 7ème supplement)*. Paris: Centre Nationale de Recherche Scientifique.

Lewin, Roger. 1993. *The origin of modern humans*. New York: Scientific American Library.

Lindly, John M., and Geoffrey A. Clark. 1990. Symbolism and modern human origins. *Current Anthropology* 31:233-61.

Little, Barbara J. 1985. A comparative analysis of spatial patterns. *American Archeology* 5:34-40.

Lumley, H. de. 1969a. *Une cabane acheuléenne dans la grotte du Lazaret, Nice*. Mémoires de la Société Préhistorique Française VII. Paris: Centre Nationale de Recherche Scientifique.

_______. 1969b. A Paleolithic camp at Nice. *Scientific American* 220:36-44.

_______. 1971. *Archaeology in the laboratory*. Paris: Service du Film de Recherche Scientifique.

_______. 1972. *La grotte de l'Hortus (Valflaunès, Hérault). Les chasseurs néanderthaliens et leur milieu de vie. Élaboration d'une chronologie du Würmien II dans le Midi Méditerrannéen*. Études Quaternaires Mémoire 1. Marseille: Laboratoire de Paléontologie Humaine et de Préhistoire.

Lyman, R. Lee. 1979. Faunal analysis: an outline of method and theory with some suggestions. *Northwest Anthropological Research Notes* 13:22-35.

_______. 1984. Bone density and differential survivorship of fossil classes. *Journal of Anthropological Archaeology* 3:259-99.

_______. 1985. Bone frequencies: differential transport, *in situ* destruction, and the MGUI. *Journal of Archaeological Science* 12:221-36.

_______. 1992. Anatomical considerations of utility curves in zooarchaeology. *Journal of Archaeological Science* 19:7-22.

Lyman, R. Lee, and Gregory L. Fox. 1989. A critical reevaluation of bone weathering as an indication of bone assemblage formation. *Journal of Archaeological Science* 16:293-317.

Marean, Curtis W., and Lillian M. Spencer. 1991. Impact of carnivore ravaging on zooarchaeological measures of element abundance. *American Antiquity* 56:645-58.

Marean, Curtis W., Lillian M. Spencer, Robert J. Blumenschine, and Salvatore Capaldo. 1992. Captive hyaena bone choice and destruction, the Schlepp Effect and Olduvai archaeofaunas. *Journal of Archaeological Science* 19:101-21.

Marshack, A. 1989. Evolution of the human capacity: the symbolic evidence. *Yearbook of Physical Anthropology* 32:1-34.

Marshall, Larry. 1989. Bone modification and "The laws of burial." In *Bone modification*, ed. Robson Bonnichsen and Marcella H. Sorg, 7-26. Orono, Maine: Center for the Study of the First Americans, University of Maine.

Mellars, Paul. 1989. Major issues in the emergence of modern humans. *Current Anthropology* 30:349-85.

Mellars, Paul, ed. 1990. *The emergence of modern humans: an archaeological perspective*. New York: Cornell University Press.

Mellars, Paul, and Christopher B. Stringer, eds. 1989. *The human revolution: behavioural and biological perspectives on the origins of modern humans*. Edinburgh: Edinburgh University Press.

Miller, George J. 1969. A study of cuts, grooves, and other marks on recent and fossil bone I. Animal tooth marks. *Tebiwa* [The Journal of the Idaho State University Museum] 12(1):20-26.

________. 1975. A study of cuts, grooves, and other marks on recent and fossil bone II. Weathering cracks, fractures, splinters, and other simlar natural phenomena. In *Lithic technology: making and using stone tools*, ed. Earl Swanston, 211-26. Chicago: Aldine.

Mondini, N. Mariana. 1995. Artiodactyl prey transport by foxes in Puna rock shelters. *Current Anthropology* 36: 520-24.

Moravské Museum. 1965. *Die Erforschung der Höhle Pod hradem 1956-1958*. Brno: Moravské Museum.

Musil, Rudolf. 1965. Die Bärenhöhle Pod hradem: die Entwicklung der Höhlenbären im letzten Glazial. In *Die Erforschung der Höhle Pod hradem 1956-1958*, 7-92. Brno: Moravské Museum.

Myers, Thomas P., Michael R. Voorhies, and R. George Corner. 1980. Spiral fractures and bone pseudotools at paleontological sites. *American Antiquity* 45:483-90.

Nichol, R. K., and C. J. Wild. 1984. "Numbers of individuals" in faunal analysis: the decay of fish bone in archaeological sites. *Journal of Archaeological Science* 11:35-51.

Nitecki, Matthew H., and Doris V. Nitecki, eds. 1994. *Origins of anatomically modern humans*. New York: Plenum Press.

Noble, William, and Iain Davidson. 1991. The evolutionary emergence of modern human behaviour: language and its archaeology. *Man* (N.S.) 26:223-53.

________. 1993. Tracing the emergence of modern human behavior: methodological pitfalls and a theoretical path. *Journal of Anthropological Archaeology* 12:121-49.

O'Connell, James F. 1987. Alyawara site structure and its archaeological implications. *American Antiquity* 52:74-108.

O'Connell, James F., Kristen Hawkes and Nicholas Blurton Jones. 1988. Hadza hunting, butchering, and bone transport and their archaeological implications. *Journal of Anthropological Research* 44:113-61.

Olsen, Sandra L. 1989. On distinguishing natural from cultural damage on archaeological antler. *Journal of Archaeological Science* 16:125-35.

Olsen, S. L., and Shipman, P. 1988. Surface modification on bone: trampling versus butchery. *Journal of Archaeological Science* 15:535-53.

Otte, Marcel, Jean-Marcel Evrard, and Alain Mathis. 1988. Interprétation d'un habitat au Paléolithique moyen. La grotte de Sclayn, Belgique. In *Upper Pleistocene prehistory of Western Eurasia*, ed. Harold L. Dibble and Anta Montet-White, 95-124. University Museum Monograph 54. University Museum Symposium Series Volume I. Philadelphia: The University Museum, University of Pennsylvania.

Paquet, P. C. and L. N. Carbyn. 1986. Wolves, *Canis lupus*, killing denning black bears, *Ursus americanus*, in the Riding Mountain National Park area. *Canadian Field Naturalist* 100:371-72.

Panoš, Vladimir. 1965. Zur Geomorphologie der Höhle Pod Hradem. In *Die Erforschung der Höhle Pod hradem 1956-1958*, 139-46. Brno: Moravské Museum.

Parkington, John. 1990. A critique of the consensus view on the age of Howieson's Poort assemblages in South Africa. In *The emergence of modern humans: an archaeological perspective*, ed. Paul Mellars, 34-55. New York: Cornell University Press.

Payne, Sebastian. 1975. Partial recovery and sample bias. In *Archaeozoological Studies*, ed. A. T. Clason, 7-17. Amsterdam: North Holland.

________. 1983. Bones from cave sites: who ate what? Problems and a case study. In *Animals and archaeology: 1. Hunters and their prey*, ed. Juliet Clutton-Brock and Caroline Grigson, 149-62. B.A.R. International Series 163. Oxford: British Archaeological Reports.

Pelišek, Josef. 1965. Die Quartärsedimente der Höhle Pod hradem in Mittleren Teil des Mährischen Karstes. In *Die Erforschung der Höhle Pod hradem 1956-1958*, 109-32. Brno: Moravské Museum.

Perkins, Dexter, Jr., and Patricia Daly. 1968. A hunters' village in Neolithic Turkey. *Scientific American* 219(5):97-106.

Peyrony, D. 1921. Les moustériens inhumaient-ils leur morts? *Bulletin de la Société Historique et Archaeologique du Périgord*, 132-39.

Rasmussen, D. Tab, ed. 1993. *The origin and evolution of humans and humanness*. Boston: Jones and Bartlett.

Richardson, P. R. K. 1980. Carnivore damage to antelope bones and its archaeological implications. *Palaeontologia africana* 23:109-25.

Rigaud, Jean-Philippe, and J.-M. Geneste. 1988. L'utilisation de l'espace dans la grotte Vaufrey. In *La grotte Vaufrey à Cenac et Saint-Julien (Dordogne): paléoenvironnements, chronologie et activités humaines*, ed. J.-Ph. Rigaud, 593-611. Mémoires de la Société Préhistorique Française XIX. Paris: Centre Nationale de Recherche Scientifique.

Rigaud, Jean-Philippe, and Jan F. Simek. 1991. Interpreting spatial patterns at the Grotte XV: a multiple-method approach. In *The interpretation of archaeological spatial patterning*, ed. Ellen M. Kroll and T. Douglas Price, 199-220. New York: Plenum Press.

Rogers, L. 1981. A bear in its lair. *Natural History* 90(10):64-70.

________. 1987. Effects of food supply and kinship on social behavior, movements, and population growth of black bears in northeastern Minnesota. *Wildlife Monographs* 97:1-72.

Rogers, L., and L. D. Mech. 1981. Interactions of wolves and black bears in northeastern Minnesota. *Journal of Mammalogy* 62:434-36.

Rolland, Nicolas, and Harold Dibble. 1990. A new synthesis of Middle Paleolithic variability. *American Antiquity* 55:480-99.

Schepartz, L. A. 1993. Language and modern human origins. *Yearbook of Physical Anthropology* 36:91-126.

Schiffer, Michael B. 1972. Archaeological context and systemic context. *American Antiquity* 37:156-65.
________. 1976. *Behavioral archaeology*. New York: Academic Press.
________. 1983. Toward the identification of formation processes. *American Antiquity* 48:675-706.
________. 1987. *Formation processes of the archaeological record*. Tucson: University of Arizona Press.
Schmid, Elisabeth. 1976. Beobachtungen an Würmeiszeitlichen Hyänenkoprolithen und zerbisen knochen. In *Problèmes ethnographiques des vestiges osseux, Thème Spécializé B*, ed. F. Poplin, 143-49. Nice: UISPP IX Congrès.
Schwarcz, H. P., R. Grün, B. Vandermeersch, O. Bar-Yosef, H. Valladas, and E. Tchernov. 1988. ESR dates for the hominid burial site of Qafzeh in Israel. *Journal of Human Evolution* 17:733-37.
Schwarcz, H. P., W. M. Buhay, R. Grün, H. Valladas, E. Tchernov, O. Bar-Yosef, and B. Vandermeersch. 1989. ESR dates of the Neanderthal site, Kebara Cave, Israel. *Journal of Archaeological Science* 16:653-59.
Scott, Louis, and Richard G. Klein. 1981. A hyena-accumulated bone assemblage from Late Holocene deposits at Deelpan, Orange Free State. *Annals of the South African Museum* 86:217-27.
Shanks, Michael, and Christoper Tilley. 1987a. *Re-constructing archaeology*. Cambridge: Cambridge University Press.
________. 1987b. *Social theory and archaeology*. Cambridge: Polity Press.
________. 1989. Archaeology into the 1990s. *Norweigian Archaeological Review* 22:1-54.
Shipman, Pat, and J. Rose. 1983. Early hominid hunting, butchering, and carcass processing behaviors: approaches to the fossil record. *Journal of Archaeological Science* 2:57-98.
Simek, Jan F., and Roy R. Larick. 1983. The recognition of multiple spatial patterns: a case study from the French Paleolithic. *Journal of Archaeological Science* 10:165-80.
Simek, Jan F., and Heather A. Price. 1990. Chronological change in Perigord lithic assemblage diversity. In *The emergence of modern humans: an archaeological perspective*, ed. Paul Mellars, 243-61. New York: Cornell University Press.
Simpson, George Gaylord. 1963. Historical science. In *The fabric of geology*, ed. Claude C. Albritton, 24-48. Massachusetts: Addison-Wesley.
________. 1970. Uniformitarianism. An inquiry into principle, theory, and method in geohistory and biohistory. In *Essays in evolution and genetics in honor of Theodosius Dobzhansky*, ed. Max K. Hecht and William C. Steere, 43-96. New York: Appleton-Century-Crofts.
Singer, R., and J. J. Wymer. 1982. *The Middle Stone Age at Klasies River Mouth*. Chicago: University of Chicago Press.
Smirnov, Yuri. 1989. Intentional human burial: Middle Paleolithic (Last Glaciation) beginnings. *Journal of World Prehistory* 3:199-233.

Smith, Fred H., Anthony B. Falsetti, and Steven M. Donnelly. 1989. Modern human origins. *Yearbook of Physical Anthropology* 32:35-68.

Solecki, Ralph S. 1955. Shanidar Cave, a Paleolithic site in northern Iraq. *Smithsonian Institution Annual Report for 1954*, 389-425. Washington, D.C.: Smithsonian Institution.

Stiner, Mary C. 1990a. The ecology of choice: procurement and transport of animal resources by Upper Pleistocene hominids in west-central Italy. Ph.D. diss., Department of Anthropology, University of New Mexico, Albuquerque.

________. 1990b. The use of mortality patterns in archaeological studies of hominid predatory adaptations. *Journal of Anthropological Archaeology* 9:305-51.

________. 1991a. Food procurement and transport by human and non-human predators. *Journal of Archaeological Science* 18:455-82.

________. 1991b. An interspecific perspective on the emergence of the modern human predatory niche. In *Human predators and prey mortality*, ed. Mary C. Stiner, 149-85. Boulder, Colorado: Westview Press.

________. 1992. Overlapping species "choice" by Italian Upper Pleistocene predators. *Current Anthropology* 33:433-51.

________. 1993. The place of hominids among predators: interspecific comparisons of food procurement and transport. In *From bones to behavior: ethnoarchaeological and experimental contributions to the interpretation of faunal remains*, ed. Jean Hudson, 38-61. Occasional Paper 21. Carbondale, Illinois: Southern Illinois University Press.

________. 1994. *Honor among thieves: a zooarchaeological study of Neandertal ecology*. Princeton, New Jersey: Princeton University Press.

Straus, L. G. 1979. Caves: a palaeoanthropological resource. *World Archaeology* 10:331-39.

________. 1985. Carnivores and cave sites in Cantabrian Spain. *Journal of Anthropological Research* 38:45-96.

________. 1990. Underground archaeology: perspectives on caves and rockshelters. In *Archaeological Method and Theory*, vol 2, ed. Michael B. Schiffer, 255-304. Tucson, Arizona: The University of Arizona Press.

Stringer, Christopher, and Peter Andrews. 1988. Genetic and fossil evidence for the origin of modern humans. *Science* 239:1263-68.

Stringer, Christopher, and Clive Gamble. 1993. *In search of the Neanderthals: solving the puzzle of human origins*. London: Thames and Hudson.

Sutcliffe, Antony J. 1970, Spotted hyena: crusher, gnawer, digester, and collector. *Nature* 227:1100-13.

________. 1973. Similarity of bones and antlers gnawed by deer to human artefacts. *Nature* 246:428-30.

________. 1977. Further notes on bones and antlers chewed by deer and other ungulates. *Deer* 4:73-82.

Sutcliffe, Antony J., D. Bramwell, A. King, and M. Walker. 1976. Cave paleontology and archaeology. In *The science of speleology*, ed. T. D. Ford and C. H. D. Cullingford, pp. 495-520. New York: Academic Press.

Thorne, A. G., and M. H. Wolpoff. 1981. Regional continuity in Australasian Pleistocene hominid evolution. *American Journal of Physical Anthropology* 55:337-49.

Torres, T. 1988. *Osos (Mammalia, Carnivora, Ursidae) del Pleistoceno de la Peninsula Ibérica.* [*Publicaciones Especiales del Boletin Geológico y Minero.* Collected and reprinted from *Boletin Geológico y Minero* XCIX (I-VI), 1988]. Madrid: Instituto Geológico y Minero de España.

Trampler, R. 1898. Die Burghöhle im Punkwathale in Mähren. *Deutsche Rundschau fur Geographie und Statistik* 20(12):529-38.

_______. 1900. Die Culturschichten in den mährischen Karsthöhlen. *Mitteilungen der K. K. Zentral-Kommission fur Erforschung und Erhaltung Kunst- und historischen Denkmale* 1:1-8.

Trinkaus, Erik. 1985. Cannibalism and burial at Krapina. *Journal of Human Evolution* 14:203-16.

Trinkaus, Erik, and Pat Shipman. 1993. *The Neandertals: changing the image of mankind.* New York: Alfred A. Knopf.

Turner, Alan. 1989. Sample selection, schlepp effects and scavenging: the implications of partial recovery for interpretations of the terrestrial mammal assemblage from Klasies River Mouth. *Journal of Archaeological Science* 16:1-11.

Valladas, H., J. L. Joron, G. Valladas, B. Ahrensburg, O. Bar-Yosef, A. Belfer-Cohen, P. Goldberg, H. Laville, L. Meignen, Y. Rak, E. Tchernov, A. M. Tillier, and B. Vandermeersch. 1987. Thermoluminescence dates for the Neanderthal burial site at Kebara in Israel. *Nature* 330:159-60.

Valladas, H., J. L. Reyss, J. L. Joron, G. Valladas, O. Bar-Yosef, and B. Vandermeersch. 1988. Thermoluminescence dating of Mousterian 'Proto-Cro-Magnon' remains from Israel and the origin of modern man. *Nature* 331:614-15.

Valoch, Karel. 1965. Die Altsteinzeitlichen Begehungen der Höhle Pod hradem. In *Die Erforschung der Höhle Pod hradem 1956-1958*, 93-106. Brno: Moravské Museum.

Valoch, K., Pelišek, J., Musil, R., Kovanda, I., and E. Opravil. 1969. Die Erforschung der Kulna Höhle bei Sloup in Mährischen Karst (Czech). *Quartär* 20:1-45.

Vandermeersch, Bernard. 1976. Les sépultures néandertaliennes. In *La préhistoire française*, vol. 1, ed. H. de Lumley, 725-27. Paris: Centre Nationale de Recherche Scientifique.

Villa, Paola. 1982. Conjoinable pieces and site formation processes. *American Antiquity* 47:276-90.

Villa, Paola, and Jean Courtin. 1983. The interpretation of stratified sites: a view from underground. *Journal of Archaeological Science* 10:267-81.

Villa, Paola, and Françoise Delpech. 1993. Magdalenian hunting and butchering activities at Grotte des Églises (French Pyrenees). Paper presented at the 58th Annual Meeting of the Society for American Archaeology, St. Louis, Missouri.

Von Endt, David W., and Donald J. Ortner. 1984. Experimental effects of bone size and temperature on bone diagenesis. *Journal of Archaeological Science* 11:247-53.

Voorhies, Michael R. 1969. *Taphonomy and population dynamics of an early Pliocene vertebrate fauna, Knox County, Nebraska.* University of Wyoming Contributions to Geology Special Paper no. 1. Wyoming: University of Wyoming, Laramie.

Whallon, Robert. 1973a. Spatial analysis of occupation floors I: the application of dimensional analysis of variance. *American Antiquity* 38:266-78.

_______. 1973b. Spatial analysis of Palaeolithic occupation areas. In *The explanation of culture change: models in prehistory*, ed. Colin Renfrew, 115-30. London: Duckworth.

_______. 1974. Spatial analysis of occupation floors II: the application of nearest neighbor analysis. *American Antiquity* 39:16-34.

White, E. M., and L. A. Hannus. 1983. Chemical weathering of bone in archaeological soils. *American Antiquity* 48:316-22.

White, Leslie A. 1959. *The evolution of culture*. New York: McGraw-Hill.

White, Randall. 1985. Thoughts on social relationships and language in hominid evolution. *Journal of Social and Personal Relationships* 2:95-115.

White, Tim D. 1992. *Prehistoric cannibalism at Mancos 5MTUMR-2346.* Princeton, New Jersey: Princeton University Press.

White, William B. 1988. *Geomorphology and hydrology of karst terrains.* New York: Oxford University Press.

Wolpoff, M. H. 1980. *Paleoanthropology*. New York: Knopf.

_______. 1985. Human evolution at the peripheries: the pattern at the eastern edge. In *Hominid evolution: past, present, and future*, ed. Philip V. Tobias, 355-65. New York: Alan R. Liss.

_______. 1989. Multiregional evolution: the fossil alternative to Eden. In *The origins of modern humans: behavioural and biological perspectives*, ed. P. Mellars and C. B. Stringer, 62-108. Edinburgh: Edinburgh University Press.

Wolpoff, M. H., Xinzhi Wu, and A. G. Thorne. 1984. Modern *Homo sapiens* origins: a general theory of hominid evolution involving the fossil evidence from east Asia. In *The origins of modern humans: a world survey of the fossil evidence*, ed. F. H. Smith and F. Spencer, 411-83. New York: Alan R. Liss.

Wolpoff, M. H., J. N. Spuhler, F. H. Smith, J. Radovcic, G. Pope, D. Frayer, R. Eckhardt, and G. Clark. 1988. Modern human origins. *Science* 241:772-73.

Wood, W. R., and D. L. Johnson. 1978. A survey of disturbance processes in archaeological site formation. In *Advances in Archaeological Method and Theory*, vol. 1, ed. M. B. Schiffer, 315-81. New York: Academic Press.

Wylie, Alison. 1982. An analogy by any other name is just as analogical: a commentary on the Gould-Watson dialogue. *Journal of Anthropological Archaeology* 1:382-401.

________. 1985. The reaction against analogy. In *Advances in Archaeological Method and Theory*, vol. 8, ed. M. B. Schiffer, 63-111. New York: Academic Press.

________. 1989a. Archaeological cables and tacking: the implications of practice for Bernstein's 'Options beyond objectivism and relativism.' *Philosophy of the Social Sciences* 19:1-8.

________. 1989b. Matters of fact and matters of interest. In *Archaeological approaches to cultural identity*, ed. Stephen Shennan, pp. 94-109. Winchester, Massachusetts: Unwin Hyman.

Wynn, Thomas. 1979. The intelligence of later Acheulean hominids. *Man* (N.S.) 14:371-91.

Yellen, John. 1977. *Archaeological approaches to the present: models for reconstructing the past*. New York: Academic Press.

Young, David E. 1989. How powerful are archaeological inferences based on experimental replication? In *Bone modification*, ed. Robson Bonnichsen and Marcella H. Sorg, 53-60. Orono, Maine: Center for the Study of the First Americans, University of Maine.

Index

O

P

Q

R

S

T

U

V

W

Y

About the Author

Robert H. Gargett is a Lecturer in Anthropology at the California State University at San Jose. He received an Honours B.A. (1st Class Honours) in Archaeology from Simon Fraser University in British Columbia (1987), and an M.A. (1989) and a Ph.D. (1994) in Anthropology from the University of California at Berkeley. His works include: "Grave shortcomings: the evidence for Neandertal burial," in *Current Anthropology*, Volume 30 (1989); "Site structure, kinship and sharing in Aboriginal Australia: implications for archaeology," in *The interpretation of archaeological spatial patterning*, edited by Ellen M. Kroll and T. Douglas Price (1991, with Brian Hayden); "What is archaeology, really?" in the *Kroeber Anthropological Society Papers*, Volume 73-74 (1992); "Pacific Island avian extinctions: the taphonomy of human predation," in *Archaeology in Oceania*, Volume 28 (1993, with Marshall I. Weisler).